REX BUNN cut his technical teeth in naval and aviation communications. His involvement in theatre has been wide ranging, including acting, directing and production management. He was introduced to the art and science of stage lighting by Bob Herbert, a Stage Manager for many years with the legendary J.C. Williamson's theatre company, and subsequently made a specialised study of the subject. As a freelance lighting designer, Rex worked his way through university, graduating in Drama with first class honours and a university medal. He now works full time with the Department of Theatre studies in the University of New England, Armidale, where he lectures on the classics and technical theatre. He is presently researching a PhD on the plays of Patrick White, while concurrently writing a book on Stage Management.

STAGE LIGHTING

Practical

REX BUNN

Currency Press • Sydney

First published in 1993 by
Currency Press Ltd
PO Box 452, Paddington NSW 2021, Australia

National Library of Australia
Cataloguing-in-Publication data
Bunn, Rex, 1938-
  Practical Stage Lighting.

  ISBN 0 86819 297 X.

  1. Stage Lighting. I. Title.

792.025

Printed by Southwood Press, Marrickville, NSW
Cover design by Trevor Hood

CONTENTS

Domestic light control
Track lighting
Components of a theatre lighting system:
*power supplies/dimmers/the lighting control
console/lanterns/lighting bars/patching systems*

Direction
Intensity
Elements of effective lighting
Ideal angle
Lighting for dance
Planning for productions

Colour filters
How filters work
Mixing colours
Colour washes:
directionality, modelling, colour combinations

Initial meeting with director
Lighting synopsis
Attendance at rehearsals
Stage areas

Lighting grid
Set design
Master plan
Elevation
The dimmer list
Colours

Preface

This manual is intended as a concise guide to the basic processes involved in theatre lighting. The manual describes a simple, effective system for lighting theatrical productions. It begins with an introduction to the basic components of theatre lighting systems: **power supplies, lighting instruments, dimmers, patching systems** and **control consoles**. The principles of effective lighting design are then outlined. Using a hypothetical production as reference, the student is taken step by step through the various stages of lighting a show, from the first meeting with the director, through rigging, patching and focusing of lights, technical rehearsals and the necessary documentation associated with lighting design, and concluding with the process of running a performance. A separate section deals with the special requirements of lighting for musicals or for dance.

In the main body of the work, technical detail is kept to a minimum, so that the student can quickly become familiar with the practical aspects of lighting. For the technically minded, and those who wish to take their study further, a technical appendix is included.

The manual is designed primarily to be used by those involved in amateur theatre, or theatre in schools, colleges or universities. However, the principles outlined are identical with those relating to lighting design in the professional theatre. The manual would therefore provide a sound basis for further study.

1

Introduction

Stage lighting began with the basic need to illuminate theatre stages so that performances could take place in a weatherproof environment and at times of day when natural light was either insufficient or absent (e.g. at night). Illumination of set and actors is still the most important requirement of a theatre lighting system, but stage lighting design has gone far beyond this fundamental need. Today, lighting design is arguably the most versatile component of the design process, contributing far more in artistic terms than mere illumination. At the same time, lighting designers have achieved a status equal to other members of the design team: annual prizes are awarded for excellence in design, and productions may become noteworthy because of the quality of their lighting, as well as for other aspects of production.

Modern technology has broadened the capabilities of theatre lighting to the point where the only limiting factor is the designer's imagination. Instruments have become more powerful and more sophisticated; control systems more complex and more flexible. In this exciting environment, decisions about the style of lighting appropriate to a production have become an important part of the overall design process.

Style in stage lighting is a vast field. At one end of the scale, we have the totally realistic or 'naturalistic' production, where every lighting state on stage simulates a naturally occurring state familiar to the audience, and where every lighting change is related to some objective on-stage event, such as a change in the apparent time of day, or the operation of on-stage lighting fixtures by actors. Accurate simulations of sunlight, moonlight, warm or cool daylight, or the various permutations of artificial light, from candles to chandeliers, are important in this style of lighting. In conjunction with scenic design, lighting may help to create beautiful stage pictures which are aesthetically pleasing in their own right.

At the other extreme is the surrealist or expressionist production, where lighting may itself become an actor in the piece, signifying or creating changes in action, pace, mood, time or location. Changes in lighting may be used to indicate the passage of time, or a particular psychological process. Unusual colours and odd angles of lighting may create a sense of fantasy. Actors may be made to appear and disappear unexpectedly. Lighting may take command of audience attention, directing it to particular parts of the stage. It may also serve as part of the scenery: by defining the area in which action is taking place, it can create the impression of a small enclosed space, or a wide vista. Visual scenic elements, as small as the bars of a cell window or as large as a city skyline, may be projected on to neutral stage structures, and whole scenes transformed at the touch of a lighting control.

The most important elements in the creation of style are the choice of lighting instrument, the modification of the colour, direction, intensity and distribution of its light, and the organisation of these factors into a cohesive design plan. The basic skills necessary to achieve this can be acquired with this manual and some practical experience. A logical next step is a more detailed study of style. A number of sources for such study are included at the end of the manual.

2

The hypothetical production

The production that will form the basis of practical work outlined in this manual I have called 'Never Say Goodbye'. No similarity is intended with any extant play.

The venue in which the play is to be produced is a small, studio-type theatre, catering for an audience of about 100. The playing area is equipped with three concentric lighting bars, and there are 40 lighting outlets adjacent to the bars.

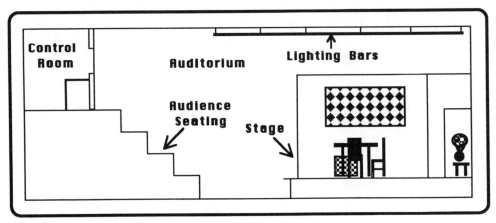

Figure 2.1 The theatre

The set contains a table and three chairs, and a sofa. There are two 'practical' doors through which entrances and exits are made.

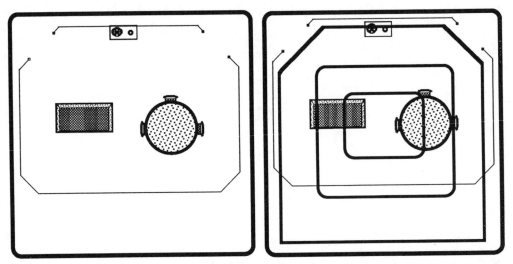

2.2 (a) Plan of the set for 'Never Say Goodbye' (b) Set with lighting bars superimposed

The play involves two characters: Tony and Sarah. The set depicts a room containing a table and three chairs, and a sofa. There is a small occasional table against the wall upstage centre, and a tapestry hanging on the wall stage right. As the play opens, Tony and Sarah are discovered in the doorway upstage right. After a brief conversation, they move into the room. There is an extensive period when the action is concentrated in the area around the table and sofa. Then Tony sits at the table and Sarah on the sofa. Each delivers a monologue, and then Tony exits through the door upstage left. Sarah is left alone on the sofa as the play concludes.

3

The hardware

The basic requirement of a theatre lighting system is that the **area** of the stage to be lit, and both the **intensity** and **colour** of the light being used, can be controlled with precision.

Most readers will be familiar with the system of variable domestic light control found in homes fairly frequently these days. A power supply comes from an outside source, and passes through a dimmer which is incorporated into a switch in the wall of a room. By turning the knob of the dimmer, the level of light in the room can be adjusted with reasonable precision.

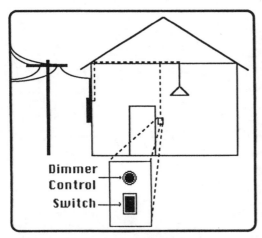

Figure 3.1 Domestic light dimmer

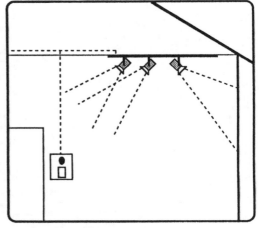

Figure 3.2 Track lighting

Track lighting, in which a number of lights may be positioned on a rail and directed to particular areas of a room, is also becoming popular.

In a theatre, a similar, though much more sophisticated, system is used. A power supply is fed through dimmers that control the intensity of a varying number of theatre lights.

There are **five** major components in a theatre lighting system:

1. a **power source**, which must be capable of supplying large amounts of power;
2. a number of **dimmer units**, each supplying one or more theatre lights;
3. a **dimmer control system**, which may be situated at some distance from the dimmers it controls. The central unit of the theatre dimmer control system is the **lighting control console**, also known as a **control desk** or **dimmer board**;
4. a variety of **theatre lights** (called **lanterns** or **luminaires**), and a flexible method of positioning them;
5. a **patching system**, used to connect lanterns to dimmers.

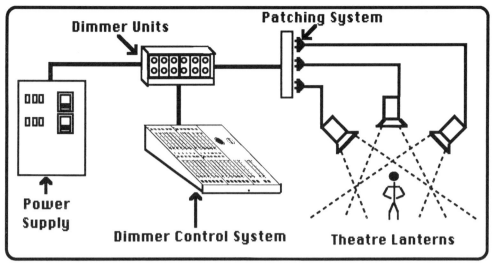

Figure 3.3 Components of a theatre lighting system

Power supplies

Small lighting systems can be purchased which will plug into a standard power point. These are useful in schools, or in portable lighting rigs for small productions. However, the operation of an adequate stage lighting system involves considerable amounts of power. The lighting system used as a reference in this manual provides for a maximum power load of just under 29 kilowatts (29 000 watts). Commercial theatres of any appreciable size require much greater amounts of power, 100 kilowatts and more being common. A three-phase system is generally necessary to supply this amount of power.

Safety note: While it is quite in order for a person without technical qualifications to rig and operate a theatre lighting system, only qualified electricians may carry out work on power supplies. The safety section contains basic advice on the safe operation of theatre lights. If electrical faults occur, do not take risks; call in a qualified electrician.

Dimmers

Dimmer units in Australia operate on 240 volts. The most commonly encountered dimmer units are rated at 2400 or 5000 watts (dimmers of lower or higher capacity are available). This means that a number of lanterns can be connected to the same dimmer, provided that the total power rating of the unit is not exceeded. A 2400 watt dimmer, for example, could be used to control two 1200 watt spotlights, or four 500 watt spotlights, or any combination of lanterns that did not exceed the total of 2400 watts.

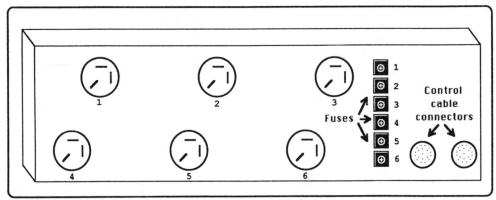

Figure 3.4 A bank of six dimmers

Dimmers usually come in **banks** or **racks** of varying numbers. Banks of six dimmers are common, and will be used as the basis of the system described in this manual.

The lighting control console

Whereas the domestic dimmer control shown in Figure 3.1 is incorporated into the dimmer/switch unit, theatre dimmers, because of the higher power they carry, usually have separate control systems which are electronically complex, and which themselves operate on lower voltages than the dimmer units they control.

Figure 3.5 shows a typical lighting console or dimmer board. This unit contains control facilities for twelve dimmers. The console has two separate **presets**, individually mastered. A **preset** is a separate set of dimmer controls, with which a lighting cue may be set up in advance (hence the term: **preset**). Each preset has its own **master** control, which governs all the dimmers in the preset simultaneously.

In the pictured unit, each of the presets is further divided into **A** and **B** circuits. Above each individual dimmer control is a three-position switch. If this switch is set to **A**, the dimmer is connected to the **A** circuit; if set to **B**, to the **B** circuit. In the central position, the dimmer is connected to both **A** and **B** circuits. Each of the **A** and **B** circuits has its own master, which means that, with some limitations, the lighting board functions as if it had four separate sets of controls.

In addition to the two presets, the lighting board has a set of **solo** or **flash buttons** (below the bottom row of dimmer controls). The **solo buttons** provide a means of bringing dimmers instantly to a predetermined level, instead of the less rapid change which occurs when using the sliding dimmer controls on the presets. These buttons are used when sudden flashes of light are required, to simulate lightning, for example, or to create a 'disco' effect. The level of all the solo buttons is set by a rotary control knob on the right-hand panel of the dimmer board. The solo

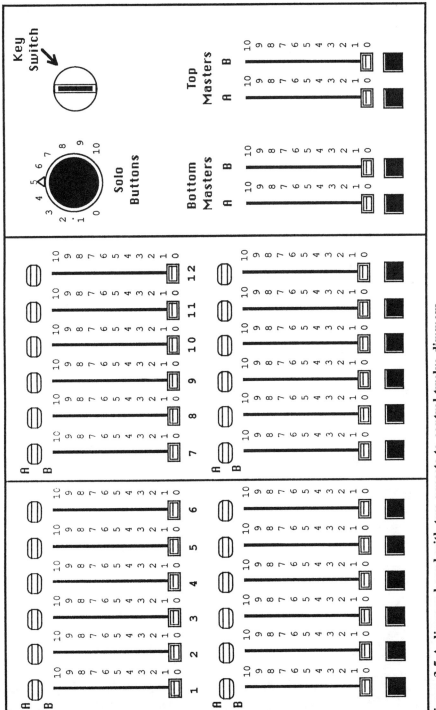

Figure 3.5 A dimmer board with two presets to control twelve dimmers

buttons operate completely independently of the presets and masters. When they are not being used, the solo buttons' control knob should be set at zero, in case the buttons are accidentally pressed while operating the presets.

Memory boards

Computerised lighting consoles, called 'memory boards', are becoming fairly common these days. Since it is fairly certain, however, that the first unit to be encountered by most readers will be a manual board, I do not propose to describe their function in this book. The progression from manual to memory board is a fairly straightforward process in any case and, as with motor cars, it is better to learn on a manual unit in case you are ever forced to 'drive' one!

The lanterns

There are three main types of theatre lantern: **floodlights**, **beamlights** and **spotlights**.

Floodlights

Floodlights are used to provide general lighting, or to light scenery, back-cloths, etc. Commonly used units range in power from 150 watts to 1200 watts. **Floodlights** do not have lenses or focusing systems. As the name suggests, the area they cover cannot be precisely controlled, though the spread of light can be limited by attaching a hood or **snoot** to the front of the unit.

Figure 3.6 Floodlights; (a) Strand 500 watt pattern 60 (b) 150 watt pattern 137

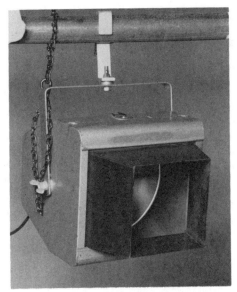

Figure 3.7 Floodlight fitted with snoot

When large areas need to be lit with floods, such as cycloramas or backcloths, banks of floodlights mounted together in a single unit, called a **batten,** may be used. Sited on the apron of the stage, battens can serve as footlights.

Figure 3.8 Flood batten

Figure 3.9 Flood battens lighting a backcloth

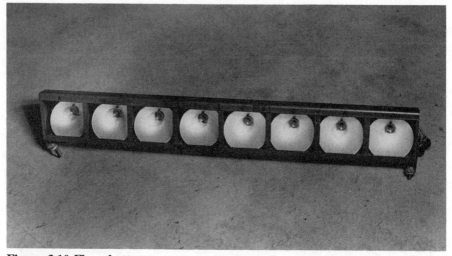

Figure 3.10 Floor batten

Beamlights

Beamlights provide a narrow beam of light, and are used to simulate a shaft of sunlight or moonlight, or the light from a street lamp, etc. **Beamlights** have focusing systems, but no lenses.

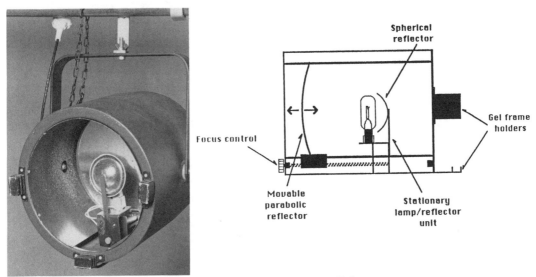

Figure 3.11 (a) Beamlight; (b) Internal structure of beamlight

A recent development of the beamlight is the **PAR can**, which uses a **PAR** lamp. The **PAR** lamp (**PAR** = **P**arabolic **A**luminised **R**eflector) is a lamp that incorporates its own parabolic reflector. It is similar in many ways to a car headlamp. The lamp is mounted in a metal cylinder to restrict light spill. **PAR cans** are simple and reliable, and produce intense, narrow beams of light. They are often seen in rock concerts, where they are used with fog machines to create colourful light shows. Theatre lighting designers working in large venues are also finding them useful. Placed in banks of several instruments and fitted with diffusion filters, they can provide strong, even colour washes over large areas.

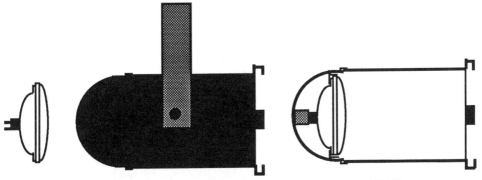

Figure 3.12 (a) PAR lamp; (b) PAR can; (c) Internal structure of PAR can

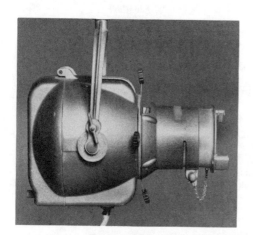

Figure 3.13 (a) and (b) Strand pattern 23 500 watt profile spot

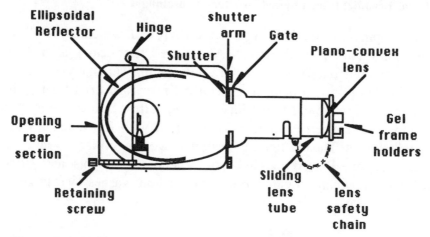

Figure 3.13 (c) Internal structure

Spotlights

Spotlights provide the most intense light, and are the most controllable of the theatre lanterns. There are a number of different types: those in more common use are **profile spots, fresnel spots, pebble convex spots,** and **follow spots.**

Profile spots usually have plano-convex lenses and sophisticated focusing and beam-shaping systems. They can be focused precisely with a hard edge to the light, and are therefore useful for projecting elements of scenery (e.g. the bars of a prison, or a crescent moon). Profile spots function in some ways like a slide projector: they have one or more adjustable focusing lenses; a set of four shutters that allow the beam of light to be shaped; and a slot in the focal plane (called the gate) into which may be inserted other shaping devices, such as an **iris** (which enables circles of varying size to be projected), a **gobo** (a metal plate with a shape cut out of it), or a **deckle** (a metal plate with holes drilled through it, to create a dappled effect, simulating light through trees).

Profile spots can also be used to provide light for acting: defocusing will soften the edge, and units with a narrow beam spread will throw a concentrated circle of light over a long distance. In large theatres, they are favoured for use on front-of-house bars, where the required length of throw may be considerable.

Figure 3.14 (a) Deckle; (b) Iris for Cantata profile; (c) Iris for pattern 23 profile

Figure 3.15 (a) Gobos; (b) Gobo in holder

Figure 3.16 Strand Cantata zoom profile spotlight

A recent development of the profile spot is the **zoom profile**, which has two adjustable lenses, controlled by sliding handles on the side of the instrument. This allows the beam spread of the lantern to be adjusted over a predetermined range. The new Strand Cantata 1200 watt profile (Fig. 3.16) comes with a standard lamp housing to which may be added any of three lens tubes, providing beam spreads of 11° to 26°, 18° to 32°, or 26° to 44° respectively.

Fresnel spots are so called because of the **fresnel** lenses with which they are equipped. A **fresnel** lens is a relatively flat lens designed to simulate the light-bending properties of a plano-convex lens. Its plane surface is textured to diffuse the light. Fresnel spots are focused by moving the position of the lamp/reflector unit in relation to the lens. They do not have the precise beam-shaping capabilities of the profile spot.

Fresnel spots provide a soft-edged beam which may be controlled in two ways: by adjusting the diameter of the beam internally, or through the use of **barn doors**, that is, shutters fitted in front of the instrument which narrow the beam of light after it leaves the lantern.

Because of their soft edge and even distribution of light, **fresnel spots** are most useful for lighting acting areas. Because of the degree of light scatter, they are most effective over short distances. In large theatres they are at their best directly over the stage.

Figure 3.17 Fresnel lens (cross-section)

Figure 3.18 Fresnel spotlights: (a) 1000 watt pattern 223; (b) 500 watt pattern 123; (c) Harmony 1000 watt fresnel

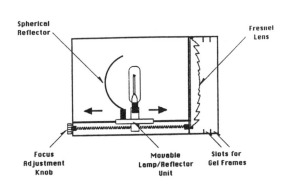

Figure 3.18 (d) Strand Minim 500 watt; (e) internal structure of the fresnel spot

Figure 3.19 (a) A set of barn doors; (b) Barn doors mounted on a spotlight

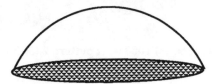

Figure 3.20 Pebble convex lens

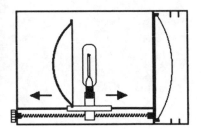

Figure 3.21 Pebble convex spotlights: (a) Strand Cantata 1200 watt PC; (b) LSC Starlette 1000 watt PC (c) Internal structure of PC spotlight

The **pebble convex spot,** commonly referred to as a '**PC**', is identical to the fresnel spot except for the structure of its lens. It has a plano-convex lens with the plane surface pebbled for diffusion. The result is an even spread of light with a slightly harder edge than the fresnel. The **PC**, like the fresnel, is favoured as an acting light, outperforming the fresnel at increased throw distances due to a lesser degree of scatter. At short range, however, the fresnel is more useful.

Figure 3.22 (a) Follow spot

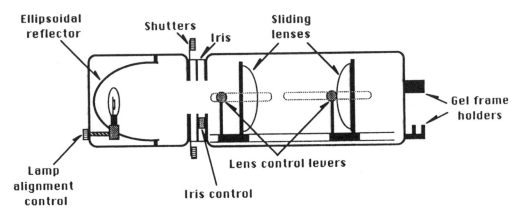

Figure 3.22 (b) Internal structure

A **follow spot** is a large **profile spot,** usually rated at 2000 watts or above, equipped with a sophisticated zoom focusing mechanism, including shutters and a built-in iris, mounted on a tripod so that it can be moved by the operator. It can throw an intense, hard-edged beam of light over a large distance. Follow spots are expensive, and it is worth remembering that any profile spotlight, suitably mounted and equipped with an iris, can be used as a follow spot.

Figure 3.23 Lantern with gel frame

Colouring lights

Theatre lanterns are provided with slots in front of the light source into which a colouring medium can be inserted. Sheets of coloured plastic film, called **gels**, are placed in frames which are then inserted into the slots. Gels come in ranges of over 100 different colours, though a selection of about 20 to 30 is sufficient for most purposes. See Chapter 5 for more details.

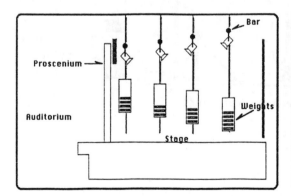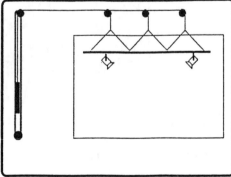

Figure 3.24 Counterweight system in proscenium arch theatre

Figure 3.25 Clamp mountings

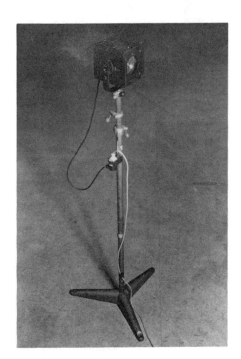

Figure 3.26 (a) Boom mounting; (b) Portable stand mount

Lighting bars

Theatre lights need to be placed in different locations to suit the particular production. The most flexible systems involve a number of lighting bars, suspended above the stage, and provided with a series of outlets into which individual lanterns are plugged. Lanterns are equipped with hook clamps which enable them to be attached to the bars quickly and securely. In proscenium arch theatres, bars are often supported by cables in a counterweight system (see Fig. 3.24). This enables the bars to be brought down to stage level for rigging, and provides flexibility in setting the height of the bar above the stage. In more open theatre spaces, the bars are usually attached to, or suspended from, the ceiling.

Lanterns may also be clamped to pieces of scenery, attached to individual brackets on a boom or ladder, or mounted on portable stands. (See Fig. 3.25, 3.26)

In Figure 3.27, a plan of the set for our hypothetical production has the lighting bars superimposed above it. The numbers in squares correspond to power outlets situated in the ceiling above the bars. This diagram will be used to assist in planning the lighting design.

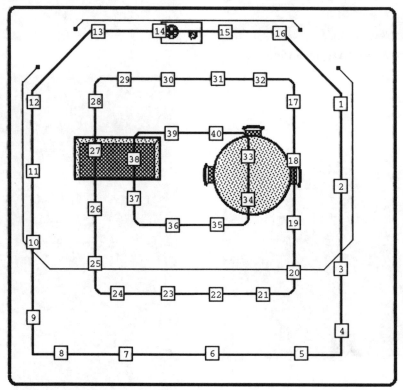

Figure 3.27 Lighting bars over the set for 'Never Say Goodbye'

The patching system

Patching systems perform two principal functions: they provide a flexible method of connecting lanterns to dimmers to suit particular productions; and they allow a large number of lanterns to be connected to a smaller number of dimmers.[1]

Most patching systems have at least two stages: lanterns are plugged into outlets, and outlets are connected to dimmers. In larger theatres, there are extra, intermediate stages, called subpatches.

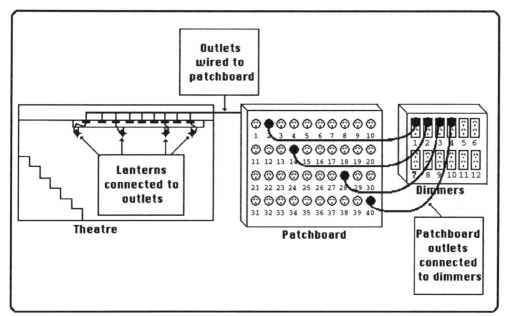

Figure 3.28 Components of the Patching System

Figure 3.28 shows the patching system for our hypothetical production. The theatre has forty outlets adjacent to the lighting bars. In the lighting control room a patchboard and short extension cords are used to connect the outlets to the twelve dimmers.

---

[1] In most theatres there are at least twice as many outlets for lanterns as there are dimmers. The 300 seat Arts Theatre at the University of New England, for example, has 101 outlets, but only 35 dimmers.

4

Lighting for acting

Lighting theatre stages for acting essentially means lighting faces. While the rest of the actor's body and stage environment must also be lit effectively, it is the actor's face which the audience watches most consistently, and it is here that deficiencies in lighting are most obvious and least tolerated by audiences.

The faces of actors on stage must be lit so that the audience sees them in sufficient detail to 'read' their expressions. To achieve this when an individual in an audience may be thirty metres or more away from the actor on stage requires a great deal of technical equipment and expertise. Facial definition is affected both by the **intensity** of the light source and by its **direction**.

Intensity

The intensity of light must be greater than is experienced outside the theatre. In a domestic situation, a single bulb of 60 – 100 watts rating is usually sufficient to provide adequate definition of a person's features. In a theatre, an actor may need to be covered by lights with a combined power of 2000 – 4000 watts, to achieve the same level of definition.

Direction

The point of origin of the light source becomes critical in a theatre. In the domestic situation, a single bulb suspended from the ceiling and reflected from the surrounding walls is sufficient. In the theatre, light for actors must come from a minimum of two separate sources, and up to eight separate light sources may be used to illuminate the same area.

Elements of effective lighting

There are no set rules for effective theatre lighting. Different lighting designers use different systems and, in the main, 'whatever looks right, is right'. However, certain standard practices and conventions of theatre lighting have become established.

The system outlined in this manual is one that I have used successfully for a number of years while designing, rigging and operating lights for a wide variety of stage productions. The system is based on principles commonly observed in the profession.

1. The ideal light for acting comes from a source placed at 45° to a horizontal plane approximately at the actor's head. Light from this source provides adequate definition, results in acceptably small shadows, and allows the light source to be concealed above the stage, or at least to be out of the audience's normal field of vision when watching the play.

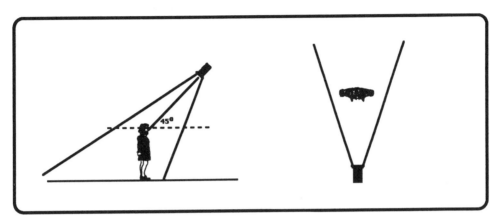

Figure 4.1 The ideal angle of lighting

2. To increase definition both of the face and of the body, light needs to come from a minimum of two sources. Once again, siting of light sources at 45° in relation to the actor is most effective.

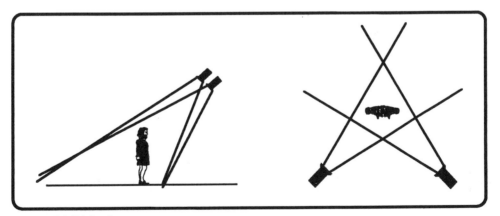

Figure 4.2 Light from two sources

3. It is also desirable for an actor to be lit from behind. Lighting from the rear has two principal effects: it helps to define the actor as a three-dimensional entity; and it separates the actor from the background.

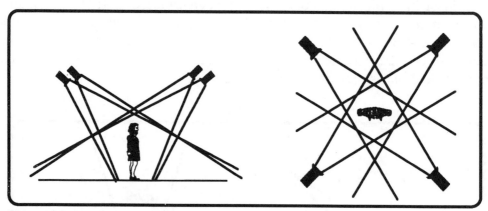

Figure 4.3 Addition of rear lighting

4. Finally, a central fill light is added, midway between the two front lights. The addition of a small amount of light from this position helps further with facial definition, and compensates for any unevenness in light distribution from the other two lanterns.

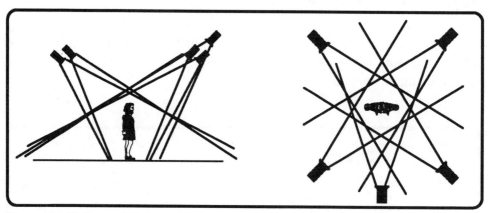

Figure 4.4 Full area lighting

For each defined stage area, this pattern of five lanterns must be repeated. The light for adjacent areas should overlap, so that an actor may move between areas without passing through a dull patch. If the available lanterns vary in their power ratings, which is usually the case, then higher powered units should be placed at 45° in front. The other three lanterns in each set may be of lower power.

For a large stage, it is obvious that a great many instruments will be needed (the number of separately lit areas times five). If there are not enough instruments to provide for this system, the number of lanterns used in each area must be reduced in a consistent way, so as to maintain uniformity.

First to go is one of the rear lights, since definition and separation can be achieved almost as well with one rear light as with two.

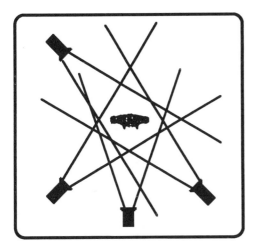

Figure 4.5 First compromise

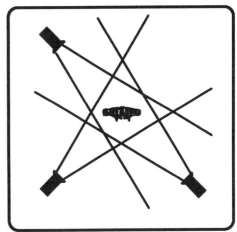

Figure 4.6 Second compromise

Next to be dispensed with is the centre fill light in front.

Finally, the second rear light may be eliminated. We are then left with two lights at 45° in front. For general acting areas, this is the minimum acceptable. If you do not have sufficient lanterns to provide two lights for each area, your lighting design will have severe limitations, and good results will be extremely difficult to achieve.

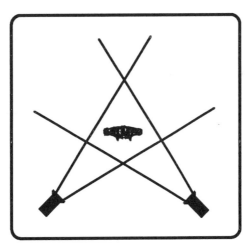

Figure 4.7 Third compromise

Lighting for dance

When lighting for dance, different principles are involved. Facial definition becomes less critical, and the movement of the body in space needs to be more effectively defined. To achieve this, vertical light and side light are very important. A certain amount of light from in front is still needed to provide sufficient illumination of the face and body from the audience's perspective. The special requirements for lighting musical and dance productions are covered in greater detail in Chapter 12.

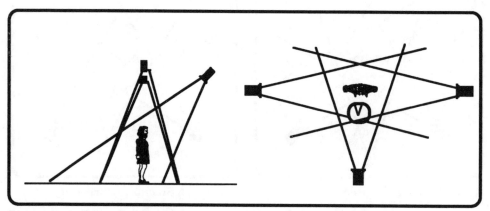

Figure 4.8 Lighting for dance

Planning for productions

When designing for an actual production, I break up the set into three different types of areas: **acting areas**, where important action takes place and/or actors deliver lines; **traffic areas**, where actors move between acting areas, but where there is no important action or dialogue; and **scenery areas**, which are never entered by actors, but where lighting is needed simply so that the audience can see scenery, furniture, props, etc. The level of lighting coverage described above is essential only for acting areas, and a lesser coverage can be tolerated for traffic areas and scenery areas. This system is explained in detail in Chapter 6. Before we proceed with the planning, however, some discussion about the use of **colour** in lights is needed.

5

Using colour in theatre lights

The colour of theatre lights may be modified for one or more of the following reasons:

1. To simulate a particular quality of natural light; e.g. sunlight, moonlight, lamplight, interior electric light, etc.
2. To enhance skin tones.
3. To enhance facial definition.
4. To enhance the colour of costumes or scenery.
5. To alter the mood or 'feeling' of a scene.
6. To convey information about a character's personality or psychological state.
7. To produce some kind of special effect.

The method of modifying colour is simple: a colour 'filter' in a metal or fibre frame is placed in front of the lighting instrument. Theatre lanterns are equipped with slots into which these filters may be inserted (see Fig. 3.23, p.26).

Colour filters

Colour filters are simply sheets of plastic film to which a colouring medium (pigment) has been added. Before the advent of modern plastics, filters were made from gelatine, hence the term 'gel', which is still used to refer to colour filters today. It should be emphasised, however, that not every coloured plastic film will be suitable for colouring theatre lights. Cheap coloured films which can be purchased from stationery shops or newsagents will not withstand the heat produced by the lamp, and will quickly burn out.

Commercially produced gels come in ranges of 100 or more different colours. A variety of brand names are in common use (e.g. Roscolux, Roscolene, Lee, Supergel, Gamcolor, Cinecolor). **Supergel**, a **Rosco** product, is a recently produced gel of high quality, and will be used as our reference system. In addition to altering colour, gels are available which diffuse the light, or alter its pattern of distribution.

Individual colours in a particular brand of gel are identified by numbers and by hue descriptions (e.g. Supergel 15 is called 'Deep Straw', while Supergel 48 is 'Rose Purple'). Confusion can arise when comparing colours from different brands of gel, since there is no consistency either in numbering systems or in colour descriptions. Supergel 66, for example, is described as 'Cool Blue'. Its equivalent in the Gamcolour range of gels is Gamcolour 720, which is called 'Light Steel Blue'.

Companies producing gels will provide sample books of gel ranges. These are called 'swatch books', and consist of small pieces of each gel colour, interleaved with cards giving the number and hue description of the gel, and in some cases a graph showing its filtering characteristics. Swatch books can usually be obtained free of charge.

Figure 5.1 Colour filter swatch books

How filters work

A colour filter works by absorbing some colours from the predominantly white light produced by the theatre lantern, while allowing others to pass through. An amber filter, for example, would absorb the blue and violet components of the light, while allowing a certain percentage of red, orange, yellow and green to pass through. (See Fig. 5.2)

Since the filter removes some of the light, the efficiency of the light source is reduced. The stronger and more pure the colour, the more light is absorbed by the filter. Filters in primary colours, for example, commonly absorb more than 80 per cent of the light. One price that must therefore be paid for modifying the colour of a theatre light is a reduction in intensity.

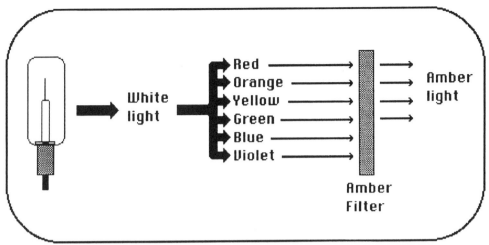

Figure 5.2 Effect of an amber filter on white light

Mixing colours

It is not necessary to purchase the full range of colours in a filter gel. For most theatres a range of 20 – 30 hues is sufficient. This does not mean, however, that the designer is limited to the colours for which gels are held, since colours can be mixed. There are three principal methods of mixing colour in light: **additive** mixing, in which two or more differently coloured lanterns are focused on the same area; **subtractive** mixing, which involves placing more than one gel in front of a single lantern; and **composite** mixing, where a gel frame is made up with pieces of differently coloured gel. Greater detail on the mixing of colours in light will be found in Appendix I.

Colour washes

The most common use of colour is in the form of a 'wash', which means that the whole of the stage will be coloured uniformly. This may be achieved by colouring all lanterns with the same filter, or by using a group of two or more lanterns, each with different coloured filters, and repeating the pattern a sufficient number of times to light the whole stage. As the light from different coloured lanterns mixes additively on stage, a uniform colour will be achieved.

Single colour washes

Washes in a single colour close to neutral can be used to produce a number of effects. A light tint of amber (Supergel 06) will give a warm colour to skin tones, similar to that of pale sunlight, or normal interior incandescent electric light. A light lavender (Supergel 54) is extremely flattering to skin tones, and is popular among lighting designers today. Its colour is close to that of normal daylight. A light tint of blue (Supergel 61 or 66) provides a cooler wash, giving skin a somewhat colder appearance. It is equivalent to an overcast exterior daylight, or a harsher interior light.

Stronger ambers (Supergel 01 or 09) provide a much warmer colour, reminiscent of strong morning or evening sunlight, or the light from paraffin lamps or candles (Supergel 11). More saturated blues (Supergel 64 or 65) or lavenders (Supergel 57A or 58A) give the appearance of moonlight.

Stronger colours

In general, the further one moves away from the lighter tints, the less realistic the scene on stage will appear. Strongly saturated gels, particularly of hues not encountered in natural light or common sources of artificial light, should therefore be avoided in productions requiring a realistic 'feel'. In musicals, productions with a strong element of fantasy, or in expressionist or surrealist plays, the imagination can be allowed greater freedom, and more saturated colours can be used to produce powerful dramatic effects.

Two-colour washes

When light is additively mixed from two different sources on a flat surface, the colours will mix to create a new colour which is a composite of the two. The colour mixture will be lighter in value (closer to white) than either of its component colours. If the two colours are complementaries (e.g. blue & amber, or green & magenta) the mixture will be white. If the surface is three-dimensional rather than flat, the colour produced will vary according to what percentage of each colour is reaching a particular part of the surface. The table which follows this section lists some commonly used two-colour washes.

In Figure 5.3, a multifaceted object is lit by two lights coloured with complementaries, blue and amber. Surface A will appear white, since it receives equal amounts of each colour. Surface B1 will receive more blue light, while surface B2 will receive more amber light. These surfaces will deviate from white towards their respective colours. Surface C1, which receives no amber light at all, will be blue, while surface C2, by the same process, will be amber. Thus five distinct colour tonings can be produced by using only two colour filters. On a multifaceted surface such as the human body, the variety of tonings is even more extensive.

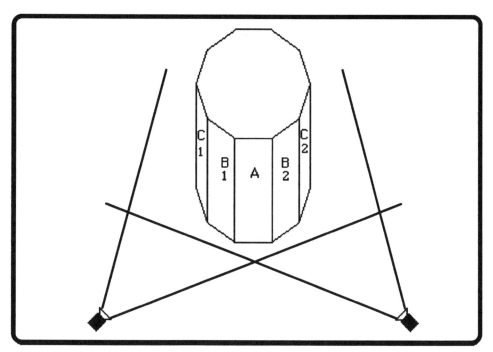

Figure 5.3 Colour mixing on a three-dimensional object. The left lantern is coloured blue and the right lantern amber.

The two-colour wash makes use of this phenomenon in two ways:

1. to enhance modelling of the human form, or objects on stage;
2. to introduce an impression of directionality (that the light is coming from a particular direction) into the stage scene, without losing definition.

Modelling

Two-colour washes enhance the **modelling** of face and form by providing more variety in tones of light and shade (as shown in Fig. 5.3). This variety of colour tonings makes the object or actor lit more interesting, and adds considerably to the audience's ability to read facial expression.

Directionality

We normally determine the direction of a light source by the pattern of light and shade it creates. The object illuminated will be in shadow on the surface that faces away from the source of light. In the theatre, however, it is not practical to create an impression of direction in this way, for two reasons. First, the implied light source will frequently be on stage, in the form of a window through which a sun or moon is shining, or a visible interior light fitting such as a table lamp or a standard lamp, whereas the actual light will be coming from in front of the stage.

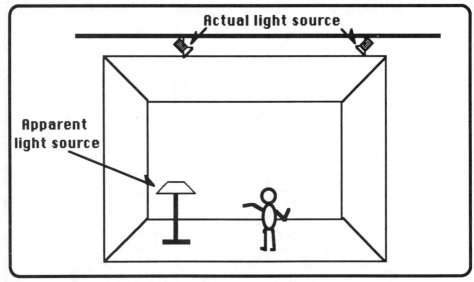

Figure 5.4 Apparent light sources on stage are an illusion

Second, to duplicate the light/shade pattern of a single, strong light source on stage will irritate an audience, since they will not be able to 'read' the expression of an actor whose face is half in shadow.

However, the *impression* of direction can be created by lighting the actor from two directions, using a warmer colour on one side and a cooler colour on the other. The warm side will be seen as reflecting the dominant source of light, while the cool side will give the impression of shadow without losing definition.

Figure 5.5 shows how a two-colour wash is used to enhance the impression of directionality in light from an onstage window. A beamlight is used backstage to throw a strong shaft of amber light (simulating sunlight) through the window, while the actor is lit by warm lights on the side close to the window, and cooler lights on the opposite side. The gels used here are Supergel 20 (medium amber) in the beamlight, Supergel 01 (light bastard amber) as the warm acting light, and Supergel 61 (mist blue) as the cool acting light. It would also be effective in this case to leave the cool acting light uncoloured, as an unfiltered light would appear cool in comparison to the amber tint. Note that while the beamlight, which is not an acting light, can be coloured with a strong amber (Supergel 20), it would not be appropriate to use such a strong colour on the actor, as this would adversely affect skin tones and costume colours.

To create the impression of moonlight, rather than sunlight, it is necessary only to alter the gels. A suitable combination would be Supergel 69 (brilliant blue) in the beamlight, Supergel 63 (pale blue) as the warm acting light and Supergel 64A (light steel blue) as the cool acting light. (N.B. Supergel 63 and 64 are both cool tints, but the lighter tint (63) will appear warmer than the other.)

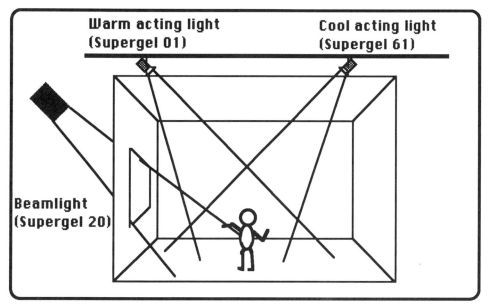

Figure 5.5 The creation of directionality

Colour combinations

Some useful two-colour washes are shown in the table below. Some of these combinations use colours that are too strong to be used in a single colour wash. In two-colour washes, more saturated colours can be used, particularly if the colours are complementary. Complementary colours mixed additively produce a wash that is lighter in value (closer to white) than either colour on its own.

First colour	Second colour	Impression
SG 05 (rose tint)	SG 61 (mist blue)	Neutral interior
SG13 (straw tint)	SG 66 (cool blue)	Normal daylight
SG 31 (salmon pink)	SG 65 (daylight blue)	Brighter daylight
SG 66 (cool blue)	SG 64A (light steel blue)	Overcast exterior daylight
SG 09 (pale gold)	SG 40 (light salmon)	Romantic evening sunlight
SG 54 (special lavender)	SG 57A (lavender)	Romantic moonlight
SG 65 (daylight blue)	SG 68 (sky blue)	Deep moonlight

Accent colours

Lights that illuminate scenery, backdrops or cycloramas can be much more deeply coloured than those used for acting, since the effect on skin tone is not a factor. When colour is used to illuminate scenery, it is good sense to use a colour that will enhance the dominant pigment colours in the scenery. Stronger colours can also be used in lights that illuminate actors from behind, since they do not affect skin colour, but are used to produce highlights, and to separate the body from its background. In naturalistic plays, the rear lights should be similar in hue, but more saturated, than the colours used to light the actors' faces.

More technical detail on the use of colour may be found in Appendix I.

6

Production: The planning stage

Initial meeting with director

The lighting designer's involvement in the production begins with his or her first meeting with the director. Prior to this meeting you should have read the script at least once, and preferably several times, so that you will have some ideas about what kind of play it is, and what type of lighting will be appropriate to it. These meetings often involve some surprises for the designer. No two directors have the same approach to lighting. Some directors are very knowledgeable about lighting design and its practicalities, and will expect to have a great deal of input to your work. Others know almost nothing about the process of theatre lighting, and will simply expect you to 'get on with the job'; you will hear from them only when they are dissatisfied with some aspect of your design. Some directors like to make lighting a feature of their production; others simply want their actors to be visible to the audience. (The *style* of the production, of course, will affect this aspect of the lighting design.)

The lighting designer must be flexible enough to accommodate the working methods of each particular director. On some productions I have developed the lighting design on the basis of only a brief conversation with the director and attendance at a couple of rehearsals. This flexibility must extend to being prepared for last minute changes. A director might discover some particular point about the play only late in the rehearsal process, and a lighting change might be necessary to communicate this development to the audience.

It is desirable that the set designer is also present at this first meeting. Since set and lighting designs should complement each other, they should be planned together so that there is an integrated approach to design. This does not always occur, however. Sometimes there is no set design (and, indeed, no set designer) and the set is simply allowed to evolve as part of the rehearsal process This is more likely to occur in studio productions, with a small, flexible acting space, where set design will principally involve the placing of furniture/rostra, etc. In a large, proscenium arch theatre, where more complex sets are the norm, a formal set design is much more likely, and necessary.

The design of sets should always take into account the position of lighting bars and the available lighting resources. As soon as you receive a copy of the set design, check it carefully to ensure that the set can be lit effectively. As lighting designer you may need to negotiate changes to the set design where they are absolutely essential for effective lighting. You might also need to hire additional equipment, if available equipment is inadequate.

The stage manager should also attend this meeting, since he or she will need to know what lighting changes are envisaged, and what part the stage manager will play in lighting during the performance. The stage manager must also schedule theatre time to allow for the hanging and focusing of lights, and for the cue-to-cue session where light levels are set.

To make your task as designer easier, there are a number of things you should try to establish at this first meeting:

1. What *style* of production is envisaged, and how will lighting contribute to that style?

2. What kind of set is envisaged? Will there be a formal set design, and if so, when will it be completed? (You must impress upon both the director and the set designer the fact that the lighting design process cannot begin until you have some idea of the set design.) In a professional production, a plan and model of the set will have been prepared before the meeting takes place, and this will be used as the basis of discussion.

3. When will the construction of the set be completed? The focusing of lights cannot be accomplished until the set is fully in place.

4. Are there to be any scene changes? If so, the crew will need to be on hand when the technician is focusing lights, so that each scene can be set up in turn, and lights focused on it.

5. What provisions are being made in the rehearsal schedule for the practical elements of lighting, such as rigging, patching, focusing, setting light levels, and establishing and rehearsing lighting cues? Inexperienced directors and stage managers frequently underestimate the amount of time taken up by the lighting process, and may be unaware that the lighting technician must have exclusive access to the theatre for whatever time is necessary to hang and focus the lights.[1]

6. How many lighting cues will there be? Has a **lighting cue synopsis** been prepared? If not, when will this be done? The synopsis, a brief description of proposed lighting cues, is essential to further planning. You should be given this synopsis as early as possible, or should prepare one yourself (see below), so that you can become familiar with the proposed running of the show.

7. Is a lighting operator to be allocated to the production? If so, how much

---

[1] It is virtually impossible to hang and focus lights while actors are rehearsing, or while crew are working on the set, and the lighting technician should not be expected to do this.

experience does the lighting operator have? Operation of lights during a performance is actually a simple procedure, and an inexperienced person can be taught to operate a lighting console after only a short period of instruction. However, it is important that the lighting operator enters the production as early as possible, and is present at all technical rehearsals, so that he or she can become familiar with the technicalities of running the show.

8. Will the stage manager call the lighting cues from prompt corner, or will the lighting operator follow a specially prepared script? Traditionally, the stage manager calls the lighting cues. However, when I operate lights I prefer to have my own script, and to follow this during the performance. With complex lighting plots, this is not always practicable, and the stage manager (or deputy stage manager) should then call the cues.

You should come away from this initial meeting with a clear idea of the style of the production, the function that lighting will play in its design concept, the approximate number of lighting cues, the extent to which the stage manager will be involved in lighting cues, and the kind of set which is envisaged. You should also have, or know when to expect:

1. a copy of the set design (if a formal design is planned);
2. a rehearsal schedule which includes provision for the hanging and focusing of lights, and for all technical rehearsals;
3. the date and time at which construction of the set will be completed;
4. a lighting synopsis;
5. name and contact details of the lighting operator, if one has been allocated.

Preparing a lighting synopsis

Either during or immediately after the first meeting with the director, a lighting synopsis should be prepared. The lighting synopsis is used as a reference by director, stage manager and lighting designer. It consists of a brief description of each lighting cue, including:

1. cue number;
2. position in script;
3. cue (dialogue or stage action that precipitates change);
4. a short description of what *actually happens* to the light on stage (*not* what the lighting operator does to *make* it happen).

A lighting synopsis for our hypothetical production is shown overleaf.

Cue no.	Script	Cue	Description
1	p1	Curtain up	Stage in darkness except for the light in doorway up right.
2	p3	Tony & Sarah move downstage	Stage lights build up to full warm interior.
3	p19	Sarah: 'You're so far away'	General stage lights reduced except for area round table.
4	p21	Tony: 'I wanted things to be right'	Light fades around table and builds around sofa.
5	p23	Sarah: 'That's how it was really'	Sarah is enveloped in blue light.
6	p23	Tony exits	Stage darkens except for blue light around Sarah.
7	p23	Play ends	Blue light fades to blackout.

Figure 6.1 Lighting synopsis

Attendance at rehearsals

Before undertaking the design of the lighting rig, the designer should attend a few rehearsals of the play, in order to determine what kinds of lighting are needed in the different parts of the set. It is better to do this later rather than earlier in the rehearsal process, for two reasons. First, early rehearsals are involved in blocking movement, a slow and tedious process for an observer (a period of hours may be spent in blocking a scene that runs for only a few minutes). Second, movement of actors tends to become more settled as the rehearsals proceed, and there is less likely to be change in that movement the closer the production is to performance. Later rehearsals are therefore a more reliable basis for lighting design. At this point, actors will normally be working in a rehearsal space, rather than the theatre, so I usually take a copy of the set design with me (or a rough sketch of the set if no design exists), so that I can mark out on the design the **acting areas, traffic areas** and **scenery areas**.

Stage areas

Acting areas are those parts of the set in which important action takes place. They need to be fully lit for acting. Probably 95 per cent or more of what takes place on stage will come under this heading. This means adequate definition is essential, particularly of faces. Wherever possible, the system of five lights per area should be used.

Traffic areas are parts of the set through which actors pass as they move between **acting areas**. They need adequate lighting, but do not need to be fully lit for acting. Doorways are a good example of this type of area. If they are used only for entrances and exits, then they can be designated 'traffic' areas, and do not need to be fully lit. A single lantern, correctly placed, may be sufficient. If, however, a piece of action takes place in a doorway, or a substantial speech is delivered there by an actor, it should be designated as an acting area, and appropriately lit. Remember, lighting for acting means lighting faces. This should be your guiding rule. If the action is such that the audience needs to see the actor's face clearly, then full lighting is essential.

Scenery areas are areas which are never entered by actors. They need to be lit, but only so that the audience can see them. Often, scenery areas do not need separate lighting, since there will always be a certain amount (all too often, an excessive amount!) of spilled and/or reflected light from the general lighting, and this may be sufficient. If extra lights are needed, then floodlights can be used, because the intensity of light need not be as great as for acting. Also, the direction of the light is less critical. In our production the area around the occasional table at upstage centre is never entered by actors. There is a vase of flowers on the table. The table needs to be seen clearly, so that it will be accepted as part of the scene. The lighting needs to be sufficient only to achieve this. The same considerations apply to the stage right wall hanging.

Specials

As well as marking out the stage for **general** lighting, any 'specials' required by the director must also be marked on the set design. **Specials** are lights used for a specific purpose in the show, and not used at any other time. They are often very tightly focused, perhaps on a single actor. In 'Never Say Goodbye', for example, there is one special: at the end of the performance, Sarah is left alone on the set, seated on the sofa. At this point, the general lights fade and Sarah is enveloped in a blue light. A single profile spot is used for this purpose, and is not used at any other stage of the production.

Specials tend to monopolise equipment as each must have its own dimmer. Because of this it is often necessary to dampen the enthusiasm of directors in their allocation of specials, so that sufficient equipment remains for general lighting. As a rough rule of thumb, no more than two out of every twelve dimmers should be allocated to specials.

Figure 6.2 shows our set design with different styles of cross-hatching used to indicate **acting areas**, **traffic areas**, and **scenery areas**. The **special** is also marked. This provides the basis on which the lighting design is constructed.

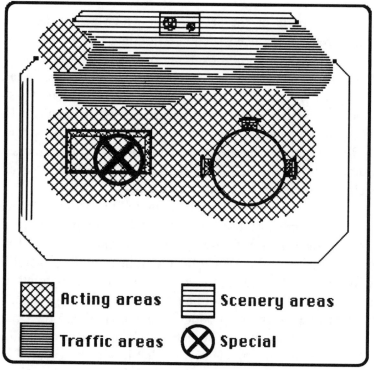

Figure 6.2 Set design with hatching and 'special' marked

7

Production: The lighting design

As outlined in Chapter 4, lighting a stage means using groups of lanterns to light particular areas. Except for extremely small designs, the whole of a set cannot be lit from one group of lanterns, so the set plan must be broken down into workable units. Before doing this, however, one important factor must be considered.

Does the design need to allow for changes of lighting during the course of performance?

If not, and if the lights are simply brought in at the beginning of the performance and taken out at the end, then the method of connecting individual lanterns to particular dimmers is not important. We could, for example, divide the set into three roughly equal sections, use a group of lights for each section, and connect each group to dimmers in any convenient fashion (remembering, of course, not to exceed the load limit of any one dimmer). There would be no need to allow for the lighting of a particular area of the stage in isolation. When the light levels were set, each dimmer would be adjusted so that there was uniform light throughout the stage. During performance, the master control would be used to raise and lower the lights over the whole of the stage.

If, however, the lighting design is required to provide for changes of lighting during performance, where some areas of the stage will be lit, and others in darkness, or in reduced light, a different approach must be used.

Our hypothetical production is of the latter kind. As we know from the lighting synopsis (Fig 6.1, p.46), the director wishes to use five separate lighting patterns during performance:

1. light in doorway upstage right only;
2. stage fully lit;
3. light principally around table;
4. light principally around sofa;
5. special on sofa only.

In this case, each of the three areas to be lit separately (doorway, table and sofa) must be covered by its own set of lights, and this group must have its own set of dimmers, so that it can be lit in isolation by the lighting operator.

The set must now be divided into areas for lighting. This process is often described as creating a lighting 'grid': the set is divided into squares, and each square is covered by a group of lanterns. Since light from theatre lanterns does not fall in squares, however, but in circles or ovals, I prefer to use the circle as the unit of division, and to overlap my circles to ensure total coverage. Using circles gives a more realistic idea of how light will fall on the set, where overlaps will occur, and in which areas undesirable spill of light can be expected. A compass is very handy at this stage of the design.

An individual group of lanterns will light a circle up to four metres, depending on the power of the lanterns, their beam width and their height above the stage. A good working diameter, however, is between two and three metres.

As lanterns are hung, they will be plugged into the nearest available outlet. Figure 7.1 shows our set design with the lighting bars and outlets drawn in.

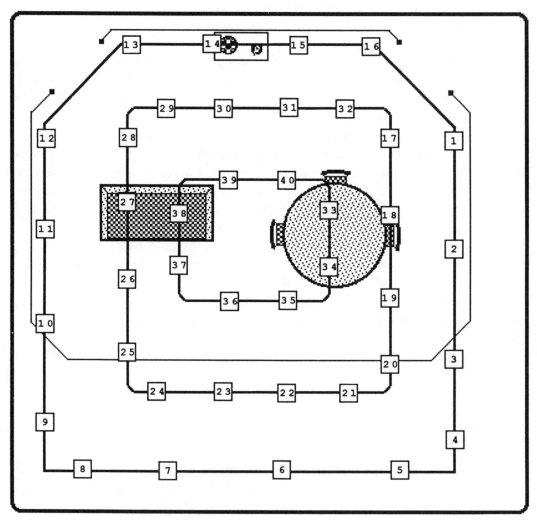

Figure 7.1 Set design showing lighting bars and outlets

Figure 7.2 shows the set design with circles indicating the important lighting areas. The positions of the lanterns that are to be hung for each area must now be sketched into the design.

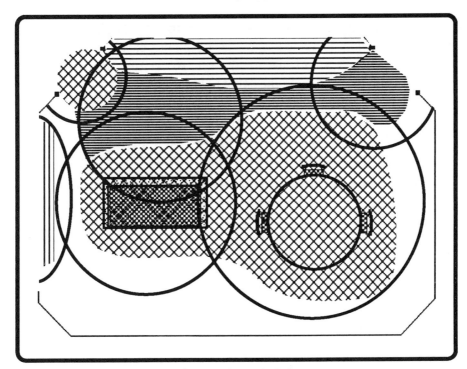

Figure 7.2 Set design with lighting sections circled

When allocating lanterns, an **elevation** or **section** (side view) of the set is useful. It allows us to check the vertical angle of a light beam that will result from hanging a lantern at any particular point on the lighting bars.

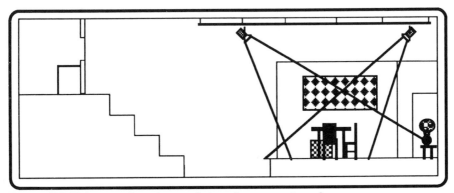

Figure 7.3 Using an elevation to determine correct angle

It will be obvious that, in a lighting design of any complexity, a great number of lanterns will be used. Master plans, which show the position of every lantern, can be extremely confusing to look at, especially for the beginner. One way of reducing this confusion is to use a series of plans, each of which shows the lanterns being used for a small area of the set. A **master** plan is then prepared, incorporating detail from each of the others. The **master** plan does not show the area lit by each lantern, but it does contain information about patching and colouring which does not appear on the individual plans. It will be used, together with the **dimmer list**, as a reference during the rigging phase of the production.

In this case, four plans, plus the master, are necessary (see Figs 7.4–7.7).

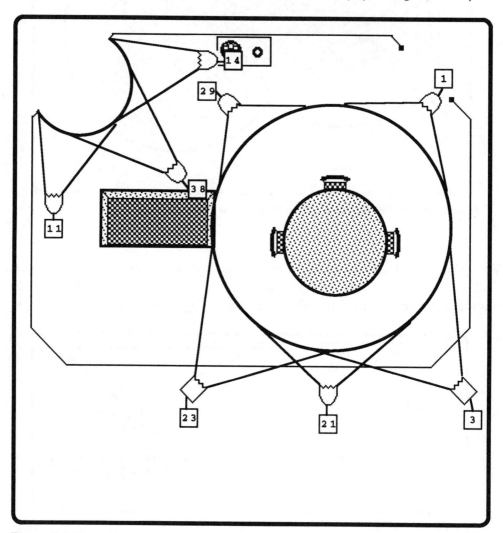

Figure 7.4 The table and the door upstage right

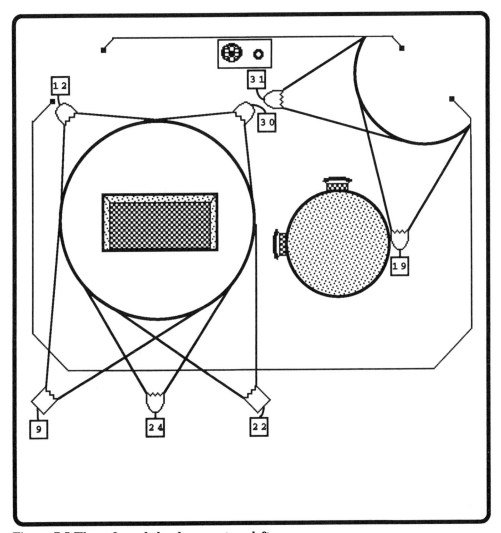

Figure 7.5 The sofa and the door upstage left

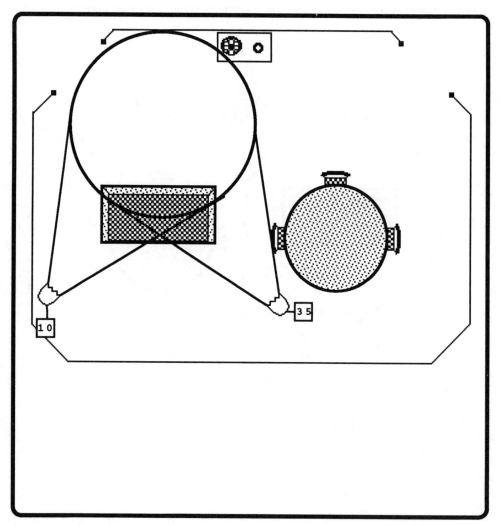

Figure 7.6 The area upstage of the sofa

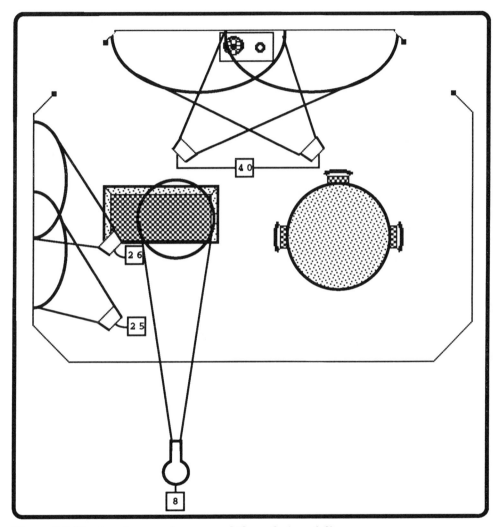

Figure 7.7 The two 'scenery' areas and the sofa 'special'

The master plan

In the master plan, the figures in **squares** refer to the number of the outlet to which each lantern is connected. The figures in **circles** are dimmer numbers. The figure **inside** the lantern icon designates the Supergel colour filter used.

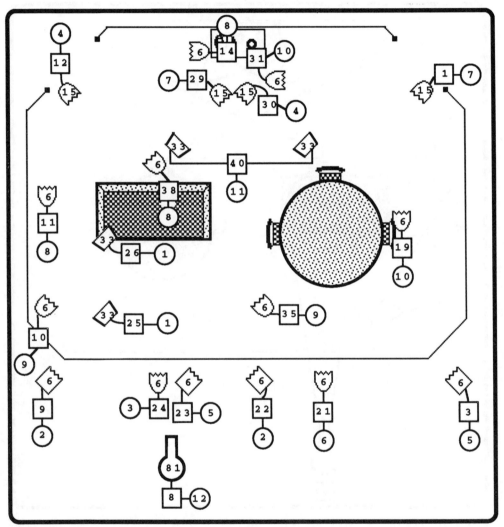

Figure 7.8 The master plan

The icons which are used to indicate lanterns also designate the **type** of lantern and its **power rating**. A key to the icons used should always be included on the master plan.

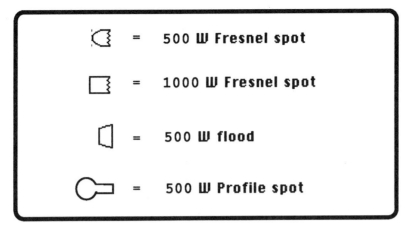

Figure 7.9 Key to the icons used in designing

	=	500 W Fresnel spot		
	=	1000 W Fresnel spot		
	=	500 W flood		
	=	500 W Profile spot		

Dimmer List

Production: _Never Say Goodbye_

Dimmer	Outlets	Lanterns	Colours	Description
1	25,26	2 x 500 W flood	33	Wall hanging (stage right)
2	9,22	2 x 1 KW Fresnel	6	Sofa (main)
3	24	1 x 500 W fresnel	6	Sofa (fill)
4	12,30	2 x 500 W fresnel	15	Sofa (rear)
5	3,23	2 x 1 KW Fresnel	6	Table (main)
6	21	1 x 500 W fresnel	6	Table (fill)
7	1,29	2 x 500 W fresnel	15	Table (rear)
8	11,14,38	3 x 500 W fresnel	6	Door, Upstage Right
9	10,35	2 x 500 W fresnel	6	Upstage of Sofa
10	19,31	2 x 500 W fresnel	6	Door, Upstage Left
11	40	2 x 500 W fresnel	33	Upstage centre
12	8	1 x 500 W profile	81	Sofa Special

Figure 7.10 The dimmer list

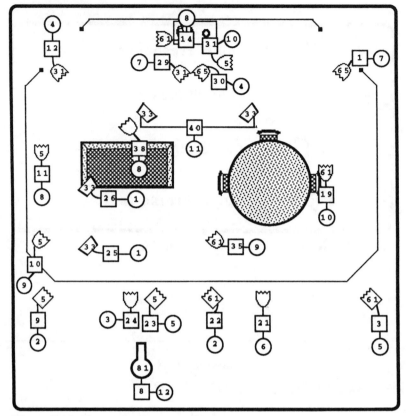

Figure 7.11 Alternative master plan

The dimmer list

When the lighting design is complete, the dimmer list is prepared. Connection of lanterns to dimmers is not arbitrary, but is designed so that the arrangement of dimmers controlling particular areas reflects to some extent the visible layout of the stage.[1] This is important for the lighting operator, who needs to be able to find dimmers relating to specific areas quickly during a performance. This list shows the outlets connected to each dimmer, the type and power rating of lanterns plugged into those outlets, the gels (colours) used in each lantern, and the area of the stage controlled by the dimmer.

The dimmer list is an important source of reference for lighting technician and operator. It is used when patching outlets into dimmers in the lighting control room; it provides a quick reference to dimmer function during lighting rehearsals;

---

[1]   For example, the dimmer controls for the **sofa lights** (2, 3 & 4) are situated to the left of those that control the **table lights** (5, 6 & 7). This duplicates the relationship between the **actual stage areas** in the operator's field of vision.

and it should be available to the operator before each performance so that the operation of lights can be checked, and any faults corrected quickly.

Dimmer List

Production: *Never Say Goodbye*

Dimmer	Outlets	Lanterns	Colours	Description
1	25,26	2 x 500 W flood	33	Wall hanging (stage right)
2	9,22	2 x 1 KW Fresnel	5/61	Sofa (main)
3	24	1 x 500 W fresnel	–	Sofa (fill)
4	12,30	2 x 500 W fresnel	31/65	Sofa (rear)
5	3,23	2 x 1 KW Fresnel	5/61	Table (main)
6	21	1 x 500 W fresnel	–	Table (fill)
7	1,29	2 x 500 W fresnel	31/65	Table (rear)
8	11,14,38	3 x 500 W fresnel	5/61/–	Door, Upstage Right
9	10,35	2 x 500 W fresnel	5/61	Upstage of Sofa
10	19,31	2 x 500 W fresnel	5/61	Door, Upstage Left
11	40	2 x 500 W fresnel	33	Upstage centre
12	8	1 x 500 W profile	81	Sofa Special

Figure 7.12 Alternative dimmer list

Colours

The colour scheme shown in Figures 7.8 and 7.10 is fairly straightforward. Frontal acting lights are coloured with Supergel 6, a pale amber reminiscent of slightly warm interior light. The rear lights are coloured with a stronger amber, Supergel 15. The floods are coloured with a light pink, Supergel 33, which indicates that there is some pink in the scenery areas that will be enhanced by this colour. The special is Supergel 81, a deep, cold blue.

An alternative colour scheme

A more ambitious colour scheme, involving a two-colour, warm-cool wash, is shown in Figures 7.11 and 7.12. Acting lights from stage right are coloured with rose tint (Supergel 5) and those from stage left with mist blue (Supergel 61). Centre fill lights are left uncoloured, and will therefore appear neutral. Rear lights are coloured with similar, but more saturated, hues: salmon pink (Supergel 31) from stage right and daylight blue (Supergel 65) from stage left. The scenery floods and the special are unaltered. Overall, this colour scheme will appear cooler than the earlier one, and will provide more variety in colour tonings.

8

Production: Rigging

With the lighting design completed, rigging of the lanterns can now proceed. Using the master plan and dimmer list as a guide, lanterns are hung on bars and plugged into the designated outlets.

Hanging

Procedure when hanging lanterns is as follows:

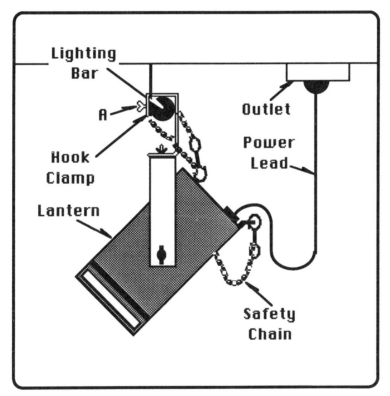

Figure 8.1 A lantern correctly hung

1. With the wing nut on the supporting bracket loosened, place the bracket over the lighting bar and allow the lantern to find its own centre of gravity. (If this is not done, uneven heat distribution through the lamp may result, which will shorten its life.)
2. With the lantern at its natural point of suspension, tighten the wing nut (A).

3. Attach safety cable or chain.
4. Ensure that the power lead is not resting against the body of the lantern.
5. If extension leads are necessary to connect a lantern to an outlet, do not allow excess lead to hang loosely, and do not wind it repeatedly around the bar to take up slack. Tape excess lead neatly to the bar with black electrical tape (*not* masking tape, which is difficult to remove).

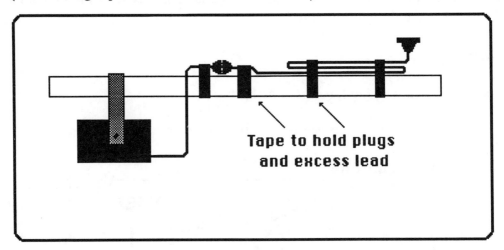

Tape to hold plugs
and excess lead

Figure 8.2 Extension lead correctly taped

When all the lanterns are in position, the next step is to patch outlets through to the appropriate dimmers.

Patching

Different theatres have different systems of patching. The system described here is fairly common in small theatres. The patchboard consists of a series of recessed plugs, each of which is connected to a single outlet over the stage. The dimmer rack has a pair of sockets on its front panel for each dimmer. Patching consists of connecting short extension leads from dimmers to patchboard. 'Piggy-back' plugs are often used so that more than one outlet can be connected to each dimmer.

Testing

Up to this point, power has not been connected to any dimmer circuit, for obvious reasons of safety. But now, with hanging and patching completed, it is time to test the lights.

   With the master on 70 per cent, each dimmer is brought in individually, and the lanterns it controls are checked to see that they correspond to the dimmer list, and

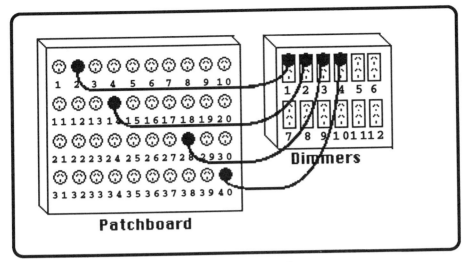

Figure 8.3 Patching system in the lighting control room

that they are functioning correctly. A note is made of non-functioning lanterns, but no action is taken at this stage.

If all dimmers and lanterns are functioning correctly, the lighting rig is ready for focusing. If there are faults, power is removed from the dimmers, and any faulty circuits are checked. The most common causes of non-function are:

1. incorrect patching (dimmers or lanterns connected to the wrong outlet);
2. careless patching (plugs not inserted properly);
3. fuses blown in dimmers;
4. lamps blown in lanterns.

The patching should be checked first. Using the dimmer list, check that the lantern is connected to the correct outlet, and the outlet to the correct dimmer, and that plugs are correctly inserted. Reconnect the power, and bring up the dimmer again.

If the circuit is still not functioning, the dimmer should be checked. Remove power from the dimmers, patch a spare lantern that is known to be functioning into the dimmer, and test again. If the dimmer does not function, check its fuse. If it does function, then the lantern is at fault.

Checking and replacing blown lamps

If a faulty lamp is suspected, disconnect power completely from the lantern at the dimmer board (switch off the control key), and unplug the lantern from the outlet. Remember, even with the dimmer control at zero, there may still be a dangerous voltage at the lamp terminals. ALWAYS DISCONNECT FIRST.

When replacing **quartz iodide** lamps, do not touch the new lamp with bare fingers. Use the plastic envelope provided, or a clean tissue. If the lamp is accidentally touched, clean it with methylated spirit and a clean tissue before inserting it. **Remember, a touched lamp, if not cleaned, is certain to burn out, probably in the middle of a performance!**

When all dimmer circuits are functioning correctly, focusing can proceed.

Focusing and colouring

Focusing is best carried out by three persons: one operating the lighting console; another on a ladder adjusting the lantern; and a third to move about the stage and serve as a model. Focusing lights without a person on stage is not impossible, but it is difficult, since light does not become visible until it strikes something, and items of scenery or the floor are not reliable guides to how the light will appear when actors are on stage.

Precautions when focusing

1. Never focus with more than 70 per cent of dimmer power. At higher levels, the lamp filament is extremely vulnerable to sudden movement, and may fracture. At up to $80 each, lamps are extremely expensive items.
2. When focusing lanterns with 'slider' focusing mechanisms, if the slider sticks, do not attempt to free it while the lamp is lit. If you do, the sudden jerk as the slider frees may fracture the filament. Remove the power, free up the slider, then restore power for focusing.
3. When focusing is completed, check that all wing nuts holding the lantern in place are tight, and that the power lead is not resting on the body of the lantern, where it can melt and cause a short circuit. **Do not use pliers or spanners to force wing nuts tight. Hand tight is sufficient.**
4. If a lamp burns out during focusing, disconnect power completely from the lantern, unplug it from the outlet, and allow it to cool before replacing the lamp.

Safety note: Even with the dimmer control at zero, there may still be a dangerous voltage at the lamp terminals. **Always disconnect first.**

Focusing theatre lights is simply a matter of pointing the lantern in the right direction and adjusting the beam width to achieve the best coverage. With profile spotlights, the softness or hardness of the beam edge can also be adjusted. (In some lanterns, the intensity level of different parts of the beam can be controlled by rotating the lamp in its housing, a control knob being provided for this purpose.) In the case of floodlights, of course, only the direction is controllable.

Figure 8.4 Stage 1 of focusing

Figure 8.5 Beam expanded to cover area

Procedure

Assuming your model to be of average height (165-170 cm) (make adjustments if your model is taller or shorter):

1. With your model standing in the centre of the area of focus, reduce the beam to its smallest area, and centre the beam on the model's chest. (see Fig 8.4)
2. Expand the area of the beam to cover the required area of focus. (see Fig 8.5)
3. Have the model walk around the perimeter of the area of focus. If light is lost on the face at any point, expand the beam to compensate. If the area being lit is on the downstage perimeter of the set, ensure also that the feet are lit at the point closest to the audience.
4. When the area is uniformly lit, tighten off all the wing nuts that hold the lantern in place.
5. If colours are being used, insert the gel and frame into the lantern (this will save another trip up the ladder).

This procedure is repeated for each lantern.

While focusing, ensure that all dimmers not in use are kept at zero. This will avoid confusing spill from lamps not being focused, and will prevent lamps not yet focused from becoming too hot to touch.

When all lanterns are focused, bring all dimmers up to 70 per cent and check that there are no dark spots between adjacent areas. If dull spots are found, adjust focus of relevant lanterns to eliminate them (in severe cases, it may be necessary to hang and focus extra lanterns to fill dull spots).

You are now ready to set the light levels for performance. If there are no lighting changes during performance, this is simply a matter of adjusting dimmers to give uniform coverage, and setting the intensity of the light to the director's satisfaction. If there are lighting changes during performance, the dimmer levels must be set for each lighting cue. This process is called the 'cue–to–cue'.

9

Production: Cue-to-cue

The cue-to-cue is sometimes referred to as a 'lighting rehearsal'. Actors are not required for this rehearsal, and you should ensure that the stage manager does not call them. They will simply stand around for a long time doing nothing and become very angry!

The objective of the cue-to-cue is to set the light levels on stage for each of the lighting cues. At this time, too, the method and timing of the transition between lighting cues is worked out. As each cue is completed, the details are entered in the state plot. The state plot records the level of every dimmer for each lighting cue, plus transition and timing between cues.

Setting light levels

With the director and lighting designer seated in the middle of the theatre, the lights for a particular cue are brought up to a satisfactory level. For areas where actors will be working, the level of light should be checked by having a model walk around the area concerned, always keeping his or her face towards the audience.[1]

When setting light levels, begin by bringing up the main frontal lights to 70 – 80 per cent. This will allow room to respond to the director's request for more or less light. When these lights are satisfactorily set, bring up the fill lights to the point where facial definition is most enhanced. Finally, bring up the rear lights until highlights appear on the model's hair and shoulders, and a clear separation from the background is evident.

As each subsequent cue is completed, the **method** and **timing** of the transition between the cues is worked out.

Method

Where a lighting cue involves bringing in a complete new preset, there are three basic methods of moving from one cue to another: **top fade**, where the new lighting cue is brought in before the previous one is faded out; **bottom fade**, where the existing lights are taken down to blackout – or near blackout – before the new cue is brought in; and **cross fade**, where the two processes overlap. A note should be made on the state plot indicating which method is to be used.

---

[1] Light levels on an empty stage are not a reliable guide to how the lighting will look with actors on stage. A model is essential.

If a cue simply involves adjustment to the preset in use, only the timing of the change need be recorded.

Timing

Timing involves setting the amount of time taken between the beginning and end of a lighting change, or between sections of the change (e.g. from full lights to blackout, and from blackout up to full lights again). Timing is usually indicated in seconds, but it may also be linked to some action on stage (e.g. the time taken for an actor to move from one point to another, or the duration of an actor's speech).

State Plot

Production: _Never Say Goodbye_   **Date:** 18/3/91

Dimmers

Cue.	Secs		1	2	3	4	5	6	7	8	9	10	11	12
1	4	V	2	2	2	2	3	3	3	8	2	1	1	-
2	Vis	∧	7	8	7	7	8	7	7	3	7	5	5	-
3	3/3	∧	0	3	2	2	"	"	"	0	0	0	0	-
4	3	X	-	8	7	7	3	2	2	-	-	-	-	-
5	5		-	"	"	"	"	"	"	-	-	5	-	-
6	3	X	-	0	0	0	0	0	0	-	-	0	-	10
7	5	V	-	-	-	-	-	-	-	-	-	-	-	0

∧ = Top Fade   V = Bottom Fade   X = Cross Fade

Figure 9.1 State plot with details entered

Using the dimmer board to set cues

In order to test the transition between cues, the following procedure should be followed during the cue-to-cue.

1. With the appropriate master on 100 per cent, set up the first cue on a preset, then record it in the state plot. Leave the cue set up, but set the master at zero.

2. Set up the next cue on the other preset, and record it.

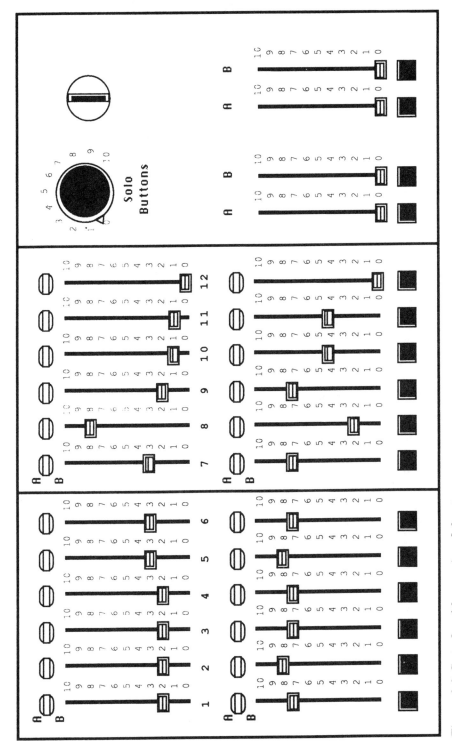

Figure 9.2 Console with cues 1 and 2 set up

3. Test the transition between cues using the preset masters. When this is done to the director's satisfaction, record details of the method and timing of the transition.

4. Zero the original preset, set up the next cue, and record it.

5. Repeat 3 and 4 until all cues are completed.[1]

Recording cues in the state plot

As each cue is completed, the levels of all dimmers used are written into the state plot. Those not used are indicated by a dash. If a dimmer level remains the same from one cue to the next, ditto marks may be used. If a dimmer is used in one cue, but not in the next, a zero should be inserted to indicate the change. Timing is recorded in seconds. If a cue takes its time from a particular event, a note is made to this effect (e.g. cue 2 for 'Never Say Goodbye' is linked to Sarah and Tony's entrance. The entry '**vis**' in the state plot indicates that the timing will be taken from a visual cue).

Figure 9.1 shows the state plot for our hypothetical production. There are seven cues altogether. Figure 9.2 shows the lighting console with the first cue set up on the top preset, and the second cue on the bottom.

A note about numbering

Alterations to cues often occur during the cue-to-cue, or at later stages in the production. Some cues may be deleted, or extra cues added. In either case, it is not necessary to renumber *all* cues. If a cue is removed, leave the numbering as it is: the missing cue number is simply ignored from then on. If extra cues are added, allocate sub-numbers to them. For example, if two extra cues are added between cues 23 and 24, they would become 23A and 23B. The original numbering thus remains undisturbed.

---

[1] Always set the master at 100 per cent when setting up a cue on a preset, so that dimmer levels will reflect the actual light level required during performance.

10

Production: Technical rehearsal

As soon as possible after the cue-to-cue, a technical rehearsal with actors should be held, to make sure that the timing of lighting cues 'works' and so that actors become familiar with the cues before the dress rehearsal. To save time, a director will often 'top-and-tail' the cues during this rehearsal, skipping the periods during which there are no lighting changes, and concentrating only on the cues themselves. When all the cues are working effectively, the lighting is ready for dress rehearsal. Before the dress, however, the lighting designer must prepare a running plot.

The running plot

State plots are prepared merely for reference purposes, and are not intended to be used during performances. For performance, a running plot is prepared. The running plot is a simplified version of the state plot, and shows only the *changes* that take place with each cue. The running plot also schedules the setting up in advance of lighting cues on the dimmer board presets. A good running plot contains every action required of the lighting operator from the beginning of the performance to the end, and must be clearly set out so that it can be followed easily amid the hubbub of an actual performance.

If a running plot is well prepared, it will be possible in an emergency for an experienced operator who has not been associated with the production to step in and, with a minimum of coaching, run the show.

The two pages of the running plot for 'Never Say Goodbye' are shown in Figures 10.1 and 10.2. Note the icons in the 'Time' boxes which indicate whether a cue is **top fade, bottom fade** or **cross fade**. The running plot begins with the setting up of the first and second cues on the two dimmer board presets (see Figure 9.2). This is done before the show commences. At curtain up, the first cue is brought in with the master. As each cue is finished, all the dimmers on that preset are zeroed, and the next cue is set up.

The broad black lines drawn across the running plot separate the plot into 'units of work', that is, actions performed in a linked series. This facilitates the use during performance of a card or ruler to cover each 'unit of work' as it is completed.

Another useful way to keep track of where you are during performance is to 'colour code' the running plot with highlighter pens. All the actions on the top preset might be higlighted in red, for example, and actions on the bottom preset highlighted in yellow. This gives operators a rapid means of confirming which preset they are working with at any particular time.

Running Plot

Production: _Never Say Goodbye_   **Page:** _1_

Cue No.	Cue.	Time.	Preset.	Dimmer changes											
				1	2	3	4	5	6	7	8	9	10	11	
	Preset 1		Top	2	2	2	2	3	3	3	8	2	1	1	
				1	2	3	4	5	6	7	8	9	10	11	
	Preset 2		Bottom	7	8	7	7	8	7	7	3	7	5	5	
1	Houselights Down	V 4	Top Master ↑	10											
2	Tony & Sarah enter	△ Vis	Bottom Master ↑	10											
			Top Master ↓	0											
				2	3	4	5	6	7						
	Preset 3		Top	3	2	2	8	7	7						
3	Sarah: "You're so far away"	△ 3	Top Master ↑	10											
		3	Bottom Master ↓	0											
				2	3	4	5	6	7						
	Preset 4		Bottom A	8	7	7	3	2	2						
	Preset 5		Bottom B	12 10											

Figure 10.1 Page 1 of the running plot

Running Plot

Production: _Never Say Goodbye_   **Page:** _2_

Cue No.	Cue.	Time.	Preset.	Dimmer changes										
4	Tony: "I wanted things to be right"	☒ 3	Bottom Master ↑	10										
			Top Master ↓	0										
5	Sarah: "That's how it was really"	5	Bottom ↑	10 / 5										
6	Tony exits	☒ 3	Bottom A Master ↓	0										
			Bottom B Master ↑	10										
7	Sarah leans back, closes eyes	☑ 5	Bottom B Master ↓	0										
		5	House lights up											

Figure 10.2 Page 2 of the running plot

As previously indicated, a lighting cue does not always involve bringing in a completely new preset. Sometimes it is simply a matter of adding one or two dimmers to the operative preset, taking them out, or changing their level. In this case the cue is effected simply by moving the appropriate dimmer controls on the preset that is functioning at the time. Cue 5 in the running plot of 'Never Say Goodbye' is an example. Here, the cue involves adding only one dimmer to the preset. Dimmer 10 is brought up to level 5. The running plot shows this.

As soon as the running plot is finished, make at least two copies of it, and keep them in separate places. If the running plot is lost, and there is no copy, a show of any complexity will not be able to be run effectively.

When the running plot is prepared, the lighting designer's job is finished. The lighting operator can now take over the running of the production. It is customary, however, for the designer to attend the dress rehearsal and the first one or two performances, to lend support to the operator, and in case minor changes to the lighting are required.

The lighting prompt script

The prompt script for the production as a whole is prepared by the stage manager. It contains exhaustive information about all aspects of the running of a performance, and normally resides in prompt corner with the stage manager or deputy. Information from the lighting cue synopsis will have been included in this script. If the stage manager (or deputy) is calling the lighting cues, the lighting operator will need only a running plot to operate during the show.[1]

If, however, the lighting operator is to run the lights unaided, a special lighting prompt script should be prepared. This is a straightforward procedure.

An unmarked copy of the script should be used. In the margin of the script, the number of the cue is printed. From this number, a line is drawn terminating in an arrow at the precise point in the script where the cue begins. If it is a simple timed cue, no further information is needed, since the running plot contains all other relevant information. If the timing is tied to some more complex event, such as an actor's movement, or the duration of a speech, this should be made clear in the prompt script. A second arrow may indicate the completion of the cue, for example, or a stage direction governing the cue may be highlighted or circled. Figures 10.3 and 10.4 show two examples from the script of 'Never Say Goodbye'.

---

[1] If lighting cues are to be called, the method of calling them should be clearly established before the technical rehearsal. A reliable method of calling cues is included in Chapter 11, **Running the show**.

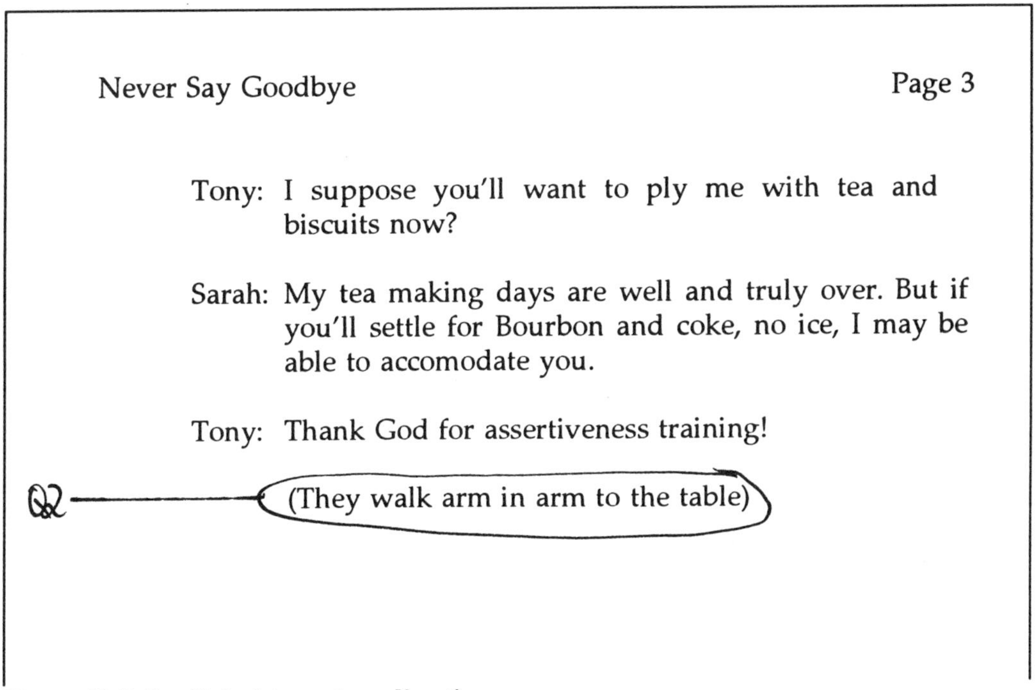

Never Say Goodbye   Page 3

Tony: I suppose you'll want to ply me with tea and biscuits now?

Sarah: My tea making days are well and truly over. But if you'll settle for Bourbon and coke, no ice, I may be able to accomodate you.

Tony: Thank God for assertiveness training!

Q2 ——————————— (They walk arm in arm to the table)

Figure 10.3 Cue linked to a stage direction

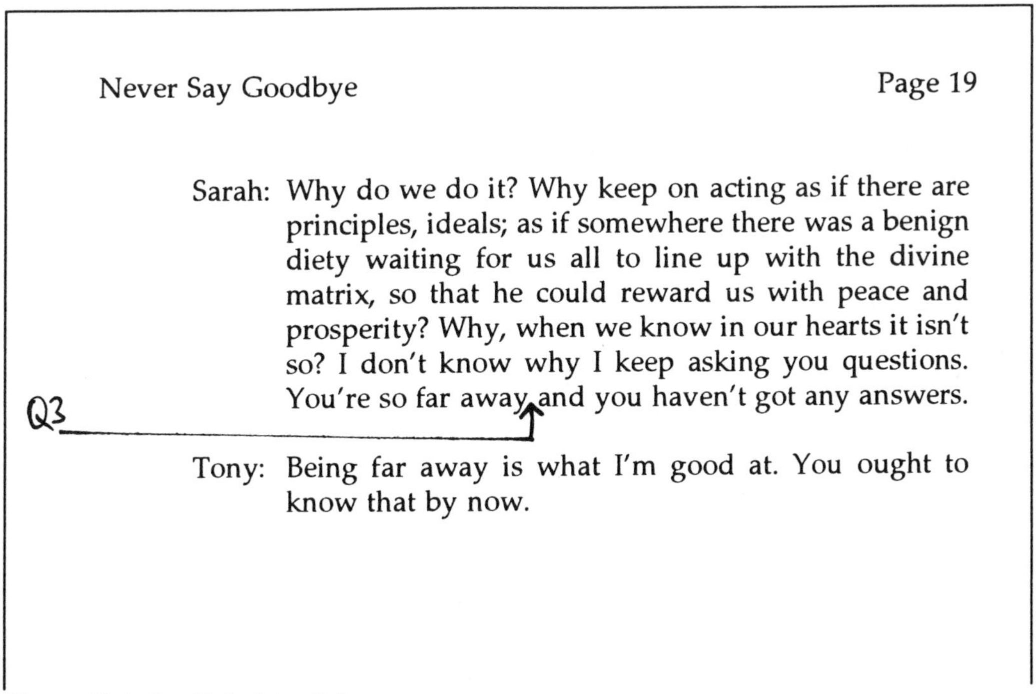

Never Say Goodbye   Page 19

Sarah: Why do we do it? Why keep on acting as if there are principles, ideals; as if somewhere there was a benign diety waiting for us all to line up with the divine matrix, so that he could reward us with peace and prosperity? Why, when we know in our hearts it isn't so? I don't know why I keep asking you questions. You're so far away and you haven't got any answers.

Q3 ——————————————

Tony: Being far away is what I'm good at. You ought to know that by now.

Figure 10.4 Cue linked to dialogue

11

Production: Running the show

For the lighting operator, the task of running a performance begins at least an hour before the curtain goes up.

Before the performance

The following routine should be carried out religiously before each performance.

1. Check all patching in the control room. Plugs often sag between performances and become disconnected. Ensure also that the power leads to lanterns on the rig are secure, and that no leads are touching the bodies of lanterns.
2. Switch on the power supply to the dimmers, turn on the dimmer board and bring all dimmers up to about 10-15 per cent to warm up the lanterns. Warming of lanterns for at least 15-20 minutes before each show is essential. If it is not done, the sudden movement of a dimmer control may cause lamps to burn out. Once the lanterns are warmed up, they will withstand sudden changes in dimmer settings.
3. While the lanterns are warming up, bring each dimmer in turn up to 70 per cent and check that all lanterns connected to it are functioning correctly.
4. As each lantern is checked for correct function, check also the condition of any gels being used. Gels tend to warp under heat, and may twist out of their frame. Darker gels, because they absorb more heat, may become blackened and useless. Another problem with gels is 'bleaching' of the colour. If any of these problems are found, replace the gel.
5. About 30 minutes before curtain-up, checking and warming of lights should be completed. It is now time to prepare the auditorium for performance. Set the level of house lights, and set up 'warmers', if required. 'Warmers' are lights used to make a set look attractive to the audience as they enter, where no curtain is being used. If the curtain is down, 'warmers' may be used to provide a blush of colour on the curtain. A simple method of providing warmers is to bring in the first lighting cue at about 20 per cent of its operating level.
6. Notify the stage manager that the lighting is set up for audience entry.
7. While the audience enters, check your running plot (pages of a running plot should be numbered clearly, so that you can check quickly that none are missing). Ensure, also, that you have copies of the dimmer list and state plot with you. (The dimmer list is essential as a form of quick reference if you need to adjust light levels on stage between cues.)
8. Set up the first one or two lighting cues on the dimmer board (if one preset is being used for 'warmers' you will be able to set up only one cue in advance).

9. Check that the 'solo buttons' control is turned off, unless it is to be used early in the performance.
10. You are now ready to run your performance. If you have done your preparation well, you can relax. A lighting operator who has taken all the necessary precautions, and who has a well-written, easily read running plot, will rarely encounter problems during a performance.

During performance.

1. As you complete each unit of work in the running plot, cover it with a ruler or card, so that you have a constant indication of where you are up to.
2. Always zero all dimmers in a preset before setting a new cue, or you may leave in dimmers from the previous cue.
3. In quiet periods, double check the next preset, and look ahead through the script and running plot to familiarise yourself with what is coming.
4. If you make a mistake with a cue, bringing it in too early or too quickly, do not go back. Chances are the audience will not notice your mistake, but if you make an obvious correction, they will certainly notice it. If you discover that you have left out a dimmer, or that there are lights on stage where they should not be, do not make sudden adjustments. Correct faults slowly, so that the audience will not notice your adjustments.
5. **Calling cues.** If lighting cues are to be called, a reliable method, consistently followed, is essential. **Brevity** and **clarity** are the important requirements. Each cue involves two calls. The first, a warning, is delivered 10-15 seconds before the change is due, and takes this form: '**Stand by, lighting cue 25**'. The second call, the actual cue, takes the form '**lighting cue 25... go**', the word '**go**' coinciding with the precise moment that the lighting change should begin. (The word '**lighting**' distinguishes the call from a *sound* cue. If sound cues are *not* being called, the word is unnecessary.)

If a series of cues occurs close together, only one '**stand by**' is necessary. For example:

'**Stand by, lighting cues 25 and 26**'
'**Cue 25, go**'
'**Cue 26, go**'

After the performance

1. Return all dimmers and masters to zero, and switch off the dimmer board.
2. Switch off the power supply to the dimmers.

3. Ensure that the control room is clean and tidy, and that all equipment has been switched off.
4. Do not leave your running plot or script in the control room. They may not be there when you come back. Take them with you and keep them in a safe place until the next performance.
5. Do not be disappointed if people do not 'notice' your lighting, and compliment you on it. In fact, be pleased! Unless extravagent effects are a feature, an audience only becomes aware of lighting when there is something wrong with it. If you do your job well, you can expect to be complimented only by people who know what is entailed in good lighting (that is, your director, the actors in the show, or other people who are involved in theatre).

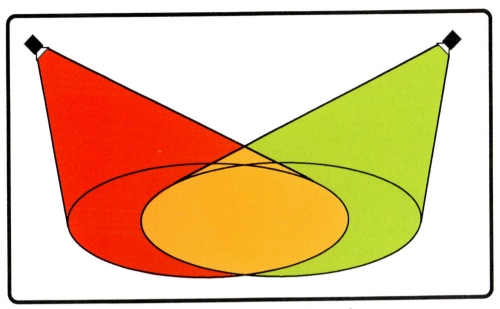

Figure A.4 Additive mixing of red and green to produce amber

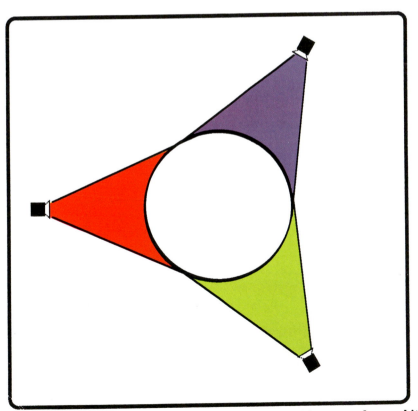

Figure A.5 Additive mixing of the three primary colours produces white

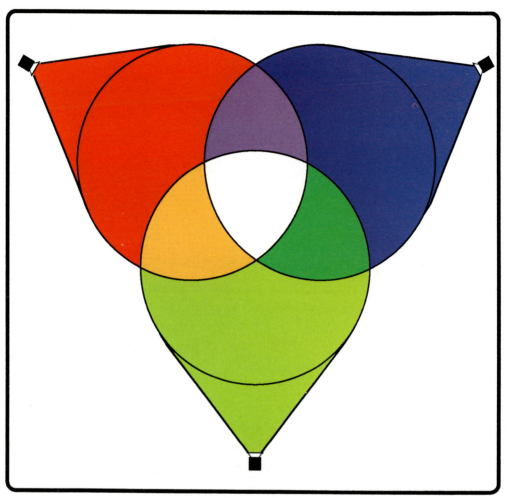

Figure A.6 The colour wheel

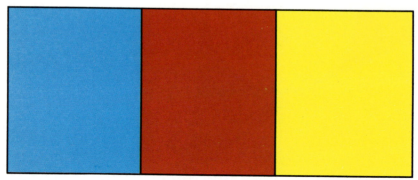

Figure A.11 The effect of simultaneous contrast on adjacent areas

Figure A.12 Cast shadows appear to be in a complementary hue to that of the light that produces them.

Figure A.13 Compatible colours in light enhance pigment colours

Figure A.14 Incompatible colour in light alters pigment colours

Figure A.15 Under certain light colours, pigment may appear black

Lighting for musicals and dance

Lighting requirements for musical and dance productions are quite different from those of drama productions. First, the audience's need to read facial expression becomes less important, while the need to define body movement adequately becomes the primary objective. As mentioned on page 30, the key change is in the direction of light. While a certain amount of frontal light is still necessary to illuminate faces, side light and vertical overhead light should dominate. Lighting from these directions accentuates bodily movement which is the central concern of this type of theatre. This means there will be a major reorganisation in the positioning of lanterns. Ideally, side light should originate from lanterns set at the same level as the dancers they light. Various kinds of vertical booms and ladders have been designed to facilitate this process.

The second requirement is a more adventurous use of colour. The need to enhance costume colours and to express mood means that strongly saturated filters will be preferred. There will also need to be rapid and diverse *changes* of colour, which requires groups of lanterns in red, blue and amber, connected to dimmers in such a way that the level of each colour can be controlled.[1] A wide range of colours can then be mixed onstage by varying dimmer levels.

A typical layout for a proscenium stage is shown in Figure 12.1. The most important lanterns are the three sets of strongly coloured side lights in red, blue and amber, hung on vertical booms at each side of the stage. Vertical light (not totally vertical, but angled slightly downstage to increase coverage and to accentuate separation of dancers) is in more strongly saturated reds and blues. Frontal light, directed straight in because its purpose is to illuminate rather than to define, is in a neutral lavender. The skycloth is lit from above and below in primaries. A ground row along the apron is also in neutral (very useful when large hats are worn on stage, and high-flown frontal light does not reach faces!).

Two follow spots are provided at the back of the theatre, spaced as wide apart as possible, to avoid 'flattening' the performer.

Proscenium arch theatres are well structured to light dance productions. In our studio some compromises are necessary. Side lighting will have to be at 45° from above, rather than at the level of the dancer, and equipment constraints demand a less extensive use of colour. Still, a workable design can be attained.

Figure 12.2 shows a design for the studio, produced within the limitations of our

---

[1] While light primaries are red, blue and green, green does not look good on human skin, and the amber mixed from red and green is not always ideal. Amber should therefore be substituted for green whenever primaries are used to light actors or dancers, rather than scenery, cycloramas, etc.

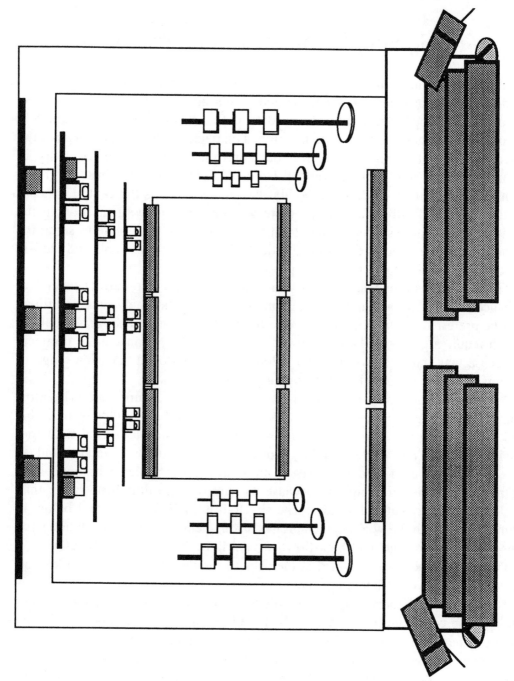

Figure 12.1 Design for a proscenium arch theatre

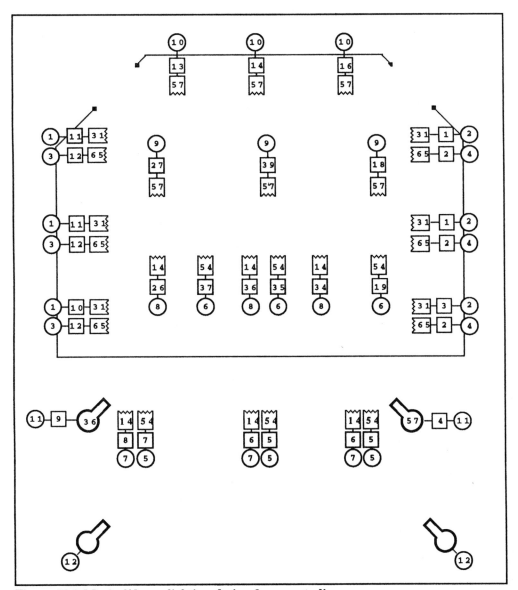

Figure 12.2 Musical/dance lighting design for our studio

twelve dimmers. The colour scheme is designed to be generally useful, and could be altered to suit particular needs, with special attention to costume colours. The lanterns used are 650 watt Strand 'Prelude' fresnels, ideal for this space. Conveniently, three of them can be supplied by one 2400 watt dimmer. Two sets of side lights provide for a red or blue stage. One set is coloured with 'salmon pink' (Supergel 31) and the other with 'daylight blue' (Supergel 65). Frontal light is also provided in two colours, a neutral 'special lavender' (Supergel 54), and an amber, 'medium straw' (Supergel 14). Rear light (vertical light is difficult with such a low ceiling) is in a more saturated lavender, called simply 'lavender' (Supergel 57). This set-up allows for three basic colour schemes, using ten dimmers and thirty instruments. A pair of profile spots in warm and cool ('medium pink' – Supergel 36, and 'light sky blue' – Supergel 67) provide a downstage centre special for solo vocal numbers (another dimmer), while two profiles equipped with irises and mounted on stands behind the audience serve as follow spots for solo dances (colours in these spots could be changed manually as required). The follow spots consume our final dimmer.

Dimmer List				
Production: _Musical Dance Variety_				
Dimmer	**Outlets**	**Lanterns**	**Colours**	**Description**
1	10, 11	3 H 650 W Fresnel	31	Side light, stage right, pink
2	1, 3	3 H 650 W Fresnel	31	Side light, stage left, pink
3	12	3 H 650 W Fresnel	65	Side light, stage right, blue
4	2	3 H 650 W Fresnel	65	Side light, stage left, blue
5	4, 6	3 H 650 W Fresnel	54	Frontal light, DS, lavender
6	19,35,37	3 H 650 W Fresnel	54	Frontal light, US, lavender
7	6, 8	3 H 650 W Fresnel	14	Frontal light, DS, amber
8	26,34,46	3 H 650 W Fresnel	14	Frontal light, US, amber
9	18,27,39	3 H 650 W Fresnel	57	Rear light, DS
10	13,14,16	3 H 650 W Fresnel	57	Rear light, US
11	4, 9	2 H 500 W Profile	36/57	DS centre solo
12	Direct	2 H 500 W Profile	As rqd	Follow spots

Figure 12.3 Dimmer list for the design in Figure 12.2

The design will require a large number of short extension leads, since there are insufficient outlets to provide one for each lantern, and two or three lanterns must frequently be connected to the same outlet. Two long extension leads will also be needed to connect the follow spot profiles directly to their dimmer, rather than via a ceiling outlet. The dimmer list for this design is shown in Figure 12.3.

To simplify running, the control console can be set up with the three basic states permanently installed. The running plot in Figure 12.4 shows how to do this. **Top preset channel A**, 'red stage', uses lateral pinks with lavender downstage. **Top preset B**, 'blue stage' substitutes blues for the pinks. **Bottom preset A**, 'amber stage', has both pinks and blues from the side (they will mix to produce lavender), and substitutes ambers for lavenders downstage. **Bottom preset B** controls the downstage solo spots and the follow spots. (During performance, the master would be left on 100 per cent, and solos or follow spots brought in by operating the individual dimmer controls.)

To run the performance, only five controls need to be operated: **top A and B masters, bottom A master, and dimmers 11 and 12 on the bottom preset**. The running plot would record only the initial set up, as in Figure 12.4, plus changes in these five controls for each cue.

Running Plot				Production: Musical/Dance Variety						Page: 1			
Cue No.	Cue	Time	Preset	Dimmer Changes									
	Red Stage		Top A	1 / 10	3 / 10	5 / 8	6 / 9	10 / 7					
	Blue Stage		Top B	2 / 10	4 / 10	5 / 8	6 / 8	9 / 7	10 / 7				
	Amber Stage		Bottom A	1 / 8	2 / 8	3 / 8	4 / 8	7 / 8	8 / 8	9 / 7	10 / 7		
	Solo/follow		Bottom B	11 / 0	12 / 0	(Master to 10)							

Figure 12.4 Running plot to set up for performance

Appendix I

Colour theory

When white light is passed through a prism, it breaks up into its basic colours. The **colour** of a beam of light is determined by its dominant wavelengths. Wavelength is measured in **nanometers**. Figure A.1 shows the position of the basic colours in the visible spectrum. For colour diagrams, please see colour section.

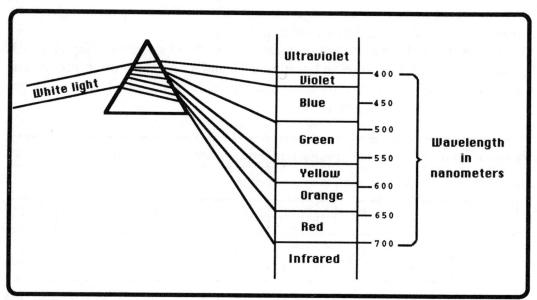

Figure A.1 The visible light spectrum

The dimensions of colour

A basic understanding of colour begins with its three dimensions: hue, value, and chroma saturation.

Hue

When we assign a name to a colour, we are usually referring to its hue. A more technical description of hue is that it refers to a colour's composition in terms of the visible light spectrum. Physicists describe hue in extremely precise terms. For lighting designers, hue usually means the colour produced by a particular gel when placed in a lighting instrument.

The major hues of the spectrum are **red, orange, yellow, green, blue** and **violet.** Variations of hue within each of these principal colours, and the virtually infinite variety of hues that can be obtained by mixing them, are given a wide range of arbitrary titles. Some imaginative examples are **bastard amber, surprise pink** and **trudy blue.**

The science of colour mixing has produced further definitions which relate to hue. **Primary** hues are those which enable a wide range of colours to be produced by mixing. Primary hues in pigment are **red, blue** and **yellow.** In light, however, they are usually described as being **red, blue** and **green.** This is slightly misleading, since the light primaries are mixtures. They are more accurately described as **red-orange, blue-violet** and **yellow-green. Secondary** hues are those which are obtained by mixing two primaries at equal intensity. **Intermediate** or **tertiary** hues are those which occur in the range between a primary and a secondary.

Value

The value of a colour is its relationship to black or white in terms of a grey scale. The number of steps in a grey scale is a matter of choice, but is commonly 7 to 10. Larger numbers of steps may be used, but serve little purpose, and may become difficult to differentiate. Figure A.2 shows a seven-step grey scale.

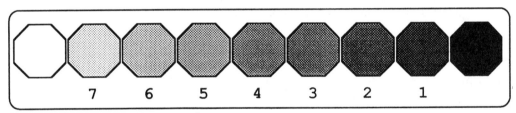

Figure A.2 A seven-step grey scale

Colours that match higher levels of the grey scale are called **tints**; those in the centre are called **tones**; those close to the black end of the scale are called **shades.** The **value** of a colour filter may be changed:

1. by adding white light from another source (value increases);
2. by inserting a filter of a complementary tint in the same frame (value decreases).

Chroma saturation

The **chroma** or **saturation** of a filter is the degree of purity of the light it produces, that is, the area of the colour spectrum which it will pass. The narrower the passband, the more saturated the colour; the wider the passband, the less saturated the colour. Thus Supergel 27 (medium red) is a **more saturated** filter than Supergel 26 (light red).

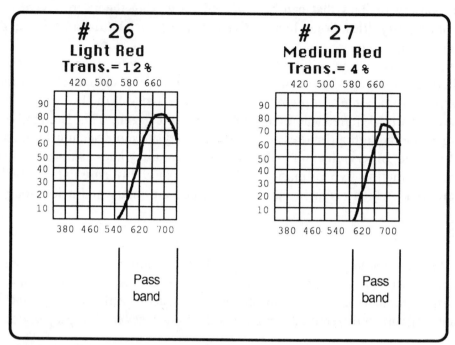

Figure A.3 Similar hues with different levels of saturation

To alter the saturation of a colour filter:

1. add more layers of the same filter (increases saturation);
2. add the colour's **complementary** from another source
   (decreases saturation).

Mixing colour

There are three principal methods of mixing colour in light: **additive mixing**; **subtractive mixing**; and **composite mixing**.

Additive mixing

In additive mixing, two or more lanterns are focused on the same area, and are coloured with gels of differing hues. If the area concerned is white, the resultant light will be a mixture of both colours. For example, a green gel in one lantern and a red gel in the other will produce an amber light (see Fig. A.4 in colour section).

Since the use of two light sources is adding together different components of the colour spectrum, the resultant colour will be lighter in value (i.e. closer to white) than either of the lanterns on its own. Additive mixing of the three light primaries (red, blue and green) from three different sources will, by the same process, produce a white light (see Fig. A.5 in colour section).

The colour wheel

Additive mixing of primaries in an overlapping pattern enables us to create a colour 'wheel'. Each pair of primaries produces a **secondary** colour which is the complementary of the third primary (complementary colours in light are those that produce white when additively mixed). As can be seen from Figure A.6 in the colour section, the following mixtures are created:

1. **Red** and **green** produce **amber**.
   Amber is the complementary of **blue**.
   Blue and **amber** produce **white**.

2. **Red** and **blue** produce **magenta**.
   Magenta is the complementary of **green**.
   Magenta and **green** produce **white**.

3. **Green** and **blue** produce **cyan**.
   Cyan is the complementary of **red**.
   Cyan and **red** produce **white**.

Where all three primaries mix in the centre the colour created is, of course, white.

Subtractive mixing

In subtractive mixing a single source of light is coloured with more than one layer of gel. The colour of the gels may be identical, in which case a more saturated version of the same hue will be produced. Alternatively, different coloured gels may be used. Each gel will remove certain components of the spectrum, and only colours which remain unabsorbed by both filters will pass through. For example, a blue-green filter will pass only the blue and green components of the spectrum; if an amber filter is added then the resultant light will be green. The amber filter absorbs the blue component; green is the only colour passed by both filters (see Fig. A.7).

Since subtractive mixing progressively eliminates more and more of the spectrum, the resultant colour will be darker in value (closer to black) than any of the gels used. Subtractive mixing thus works best with filters of light value (i.e. tints). Some combinations of gels may, in fact, absorb all of the spectrum colours, in which case no light would pass through at all (see Fig. A.8).

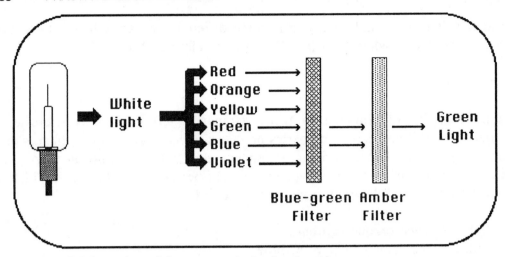

Figure A.7 Subtractive mixing progressively absorbs colours

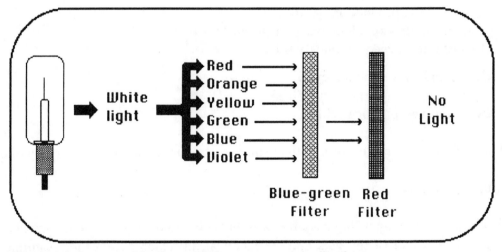

Figure A.8 Some combinations of filters may remove all colour

Composite mixing

In composite mixing a gel frame is made up using pieces of different coloured gel. Mixing of the different hues will occur on the surface lit by the lantern, and the result will be similar to additively mixing the colours from different sources (see Fig. A.9)

Another composite method is to cut a hole in a piece of gel whose colour is too strong, thus allowing some white light to pass through. The white light will mix with the coloured light and the result will be lighter in value than that produced by an uncut piece of the same gel (see Fig. A.10).

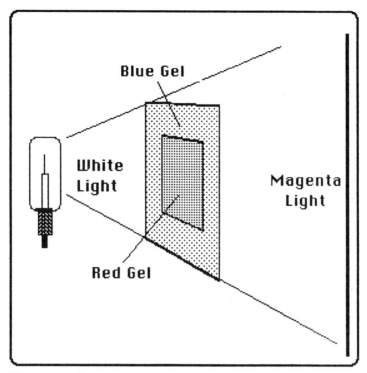

Figure A.9 Composite mixing of blue and red

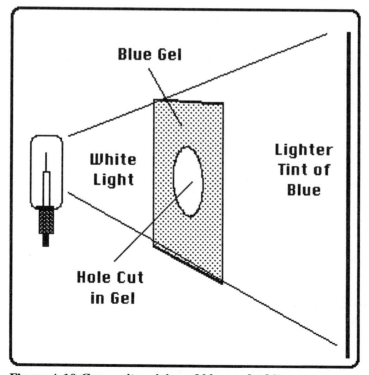

Figure A.10 Composite mixing of blue and white

The hue of a light may be changed:

1. by changing the gel;
2. by mixing colours additively, subtractively or compositely;
3. by changing the **intensity** of the light source (see **Amber Shift**, below).

Light colour and intensity

Colour temperature is the physicist's method of measuring the colour composition of a beam of light. The higher the colour temperature (measured in degrees Kelvin), the cooler (more bluish) the light; the lower the temperature, the warmer (more reddish) the light.

Theatre lamps are designed to produce light with a colour temperature of approximately 3200° Kelvin. This is a fairly cool white light. However, as the intensity of a lamp is reduced by decreasing the dimmer level, the light emitted will become progressively more yellow. This phenomenon, called **amber shift**, means that any colour filter will have amber added to it subtractively as the intensity of the lantern is reduced. If lights are to be used at levels well below maximum intensity, amber shift becomes a significant factor, and must be allowed for in the choice of filter colour.

To retain the colour of white light at low intensities, the addition of a light blue filter (e.g. Supergel 61 or 66) will help to counteract the effects of amber shift.

Some interesting facts about colour perception

Colour overload

The way in which the human eye perceives colour produces some interesting effects. The ability of the retina to perceive a colour diminishes progressively as that colour is observed. When the colour is removed or changed, the retina takes some time to recover. This phenomenon, called colour fatigue or colour overload, produces a number of effects that must be taken into consideration in lighting design, particularly where heavily saturated colours are to be used.

Diminishing intensity

As the retina becomes saturated with a colour, the ability to 'see' the colour diminishes, so the colour appears to fade.

After-image

When a colour is removed, the observer will 'see' an after-image of the coloured area for some time, because of the slow recovery of the retina. This after-image will not be in the original colour, however, but its complementary. So, if an object is lit with **red**, the after-image will be **blue-green**, the complementary of red.

Successive contrast

Successive contrast describes the effect that occurs when one colour is followed by another colour, rather than white light. Since the after-image will be in the complementary of the first colour, the effect will be to add that complementary to the second colour. Thus, if **green** is followed by **blue,** then green's complementary – **magenta** – will be added to the blue. The colour as perceived will be **blue** plus **magenta**. As the eye recovers, the magenta component will gradually fade, and the blue will be perceived in its true colour.

Simultaneous contrast

If two differently coloured areas are observed in close proximity, each adjacent area will be 'pushed' towards the complementary of the other. Thus, if a **red** area is adjacent to a **blue** area, then the **red** will have **amber** (the complementary of blue) added to it and the **blue** will have **cyan** (the complementary of red) added to it. Another way of expressing this phenomenon is to say that the eye **subtracts** adjacent colours from each other (**adding** the complementary is similar to **subtracting** the original colour).

Figure A.11 shows this effect (see Fig. A.11 in colour section). Cover the two outer panels and look at the centre, red panel. First uncover the left panel. The red is pushed towards the complement of blue (amber) and appears more orange. Cover the left panel, wait a few seconds, then uncover the right panel. The red is pushed towards the complement of yellow (violet). With all panels uncovered, the effect of each colour on its neighbour can be seen.

Colour of shadows

Shadows thrown by an object in coloured light will be perceived in the original colour's complementary. Thus an object lit with a **blue** light will appear to have an **amber** shadow; an object lit by **green** light will throw a **magenta** shadow, and so on. (The phenomenon is more easily demonstrated if some white light is added to the shadow from a second lantern. This is more indicative of stage conditions, where there is almost always some spilled light from elsewhere to enhance the colour of shadows.) See Figure A.12 in colour section.

Effect of colour on pigment

When coloured light strikes pigment in a coloured surface, such as painted scenery, costumes, or even an actor's make-up, the effect is similar to subtractive mixing. The pigment will absorb some colours from the light, while reflecting others. If this phenomenon is not considered carefully at the design stage, some odd effects can result.

If the light contains all of the colours in the pigment, then the effect will be to enhance the pigment colour. Blue light on a blue dress, for example, will simply make the colour of the dress appear richer (see Fig. A.13 in colour section).

If some of the colours in the pigment are missing from the light, then the colour will be modified. The modification may be predictable. A blue-green dress under a green light will appear dark green, since the blue component is missing, and the green is enhanced (see Fig. A.14 in colour section). Some colour combinations, however, produce unexpected and dramatic results. A yellow costume under a violet light, for example, will turn scarlet.

As in subtractive mixing, some combinations of light and pigment will result in no light being reflected, and the pigment will appear black or grey (see Fig. A.15 in colour section).

This ability of light colour to change the colour of pigment need not always be a disadvantage. It can, in fact, be a useful design tool. It may be of value to a director, for example, to have an actress enter in what appears to be a black dress, and have it suddenly turn vibrant blue as the colour of the lights changes. Changes in the colour of scenery may also be dramatically useful.

One area where great care should be exercised is in the matching of make-up to light. The subtractive effect of make-up pigments may produce some startling effects. Under a blue light, for instance, carefully rouged cheeks may appear hollow as the pigment turns to a deep purplish black. Amber light will turn lavender eye-shadow dark brown, and bright red lipstick will be rendered grey under a blue light, or black under a blue-green light.

There is no infallible method of predicting colour changes in onstage pigments. The only safe means of testing is to duplicate the onstage conditions. A piece of timber can be painted with the proposed scene paint and subjected to light coloured with a variety of filters. Costumes can be tested in the same way. Scene paints and costumes should be tested as early as possible in the rehearsal period, since colours in the lighting design may need to be changed to suit them. (A costume may sometimes be rejected because it does not suit the light, but more often the reverse will apply.)

Since make-up will most commonly be chosen to suit the lighting, it can be tested at a late rehearsal, after the lights have been set. The technical rehearsal is a good time to do this. The dress rehearsal is probably too late, though last-minute changes are always possible. If facilities allow, some spare theatre lanterns, coloured with the dominant filters in the design, may be set up before a mirror in the dressing room so that actors can check make-up before they go onstage for each performance.

Appendix II

Lamps, reflectors and lenses

The most important components of a theatre lantern are the **lamp**, the **lens** and the **reflector**. In this section we will examine each of these components in greater detail.

Lamps

The most widely used light source in stage lanterns is the **incandescent lamp**. This consists of a **clear glass envelope**, filled with an **inert gas**, and containing a **tungsten filament**. The glass envelope terminates in a **metal or ceramic base**, which also contains the lamp's electrical connections.

Many different types of lamp are commonly encountered. The features that distinguish the different types are:

1. the shape of the glass envelope;
2. the type of glass from which it is made;
3. the type of gas it contains;
4. the structure of the filament;
5. the structure of the base connector.

A theatre of any vintage will contain a good variety of different instruments, some old, some new. Lamps continue to be supplied for the older type instruments, so the lamp complement of the average theatre can be expected to provide an historical overview of lamp design through the last several decades.

The two lamps in Figure A.16 are little altered from the original design by Edison, the inventor of the incandescent light bulb. On the left is a 150 watt, pear-shaped lamp. The base is an **ES** (Edison Screw) type. On the right is the 500 watt version. The base, a larger copy of that on the left, is known as the **GES** (Goliath Edison Screw). These two lamps may be found in the **pattern 137** and **pattern 60 floodlights** respectively (see Figure 3.6, p.16).

The lamps in Figure A.17 are known as **PF** (prefocus) types. They are distinguished by their flat filament and the **prefocus** base which is designed so that it can be inserted in only one way. Correct insertion lines up the filament in parallel with the reflector. The lamp on the left, a **T1**, rated at 500 watts, is used in a variety of older lanterns, notably the **pattern 23 profile spotlight** (Fig. 3.13, p.20) and the **pattern 123 fresnel spotlight** (Fig. 3.18b, p.23). A 1000 watt version, the **T6**, identical in appearance except for its filament, is used in the larger **pattern 223 fresnel** (Fig. 3.18(a), p.23).

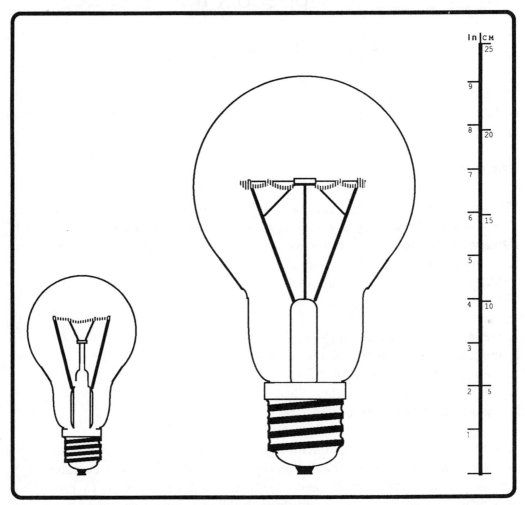

Figure A.16 Edison type lamps, used in floodlights

The disadvantages of the older type lamps are their large size, which means that everything else about the instrument that contains them must be correspondingly large, and their limited life, due to the fact that the heated filament throws off particles of tungsten, which are deposited on the inside of the glass envelope. This has two effects: first, the glass becomes progressively more opaque, reducing the level of light emitted; at the same time, the filament becomes progressively thinner, eventually burning out much earlier than it otherwise would.

All these problems were solved by the advent of the **tungsten halogen** lamp (also known as '**quartz halogen**' or '**quartz iodide**'). The tungsten halogen lamp contains a 'halogen' gas, which causes the radiated particles of tungsten to be redeposited on the filament, thus lengthening the life of the lamp. This process requires a greater concentration of heat, so the glass envelope has been made

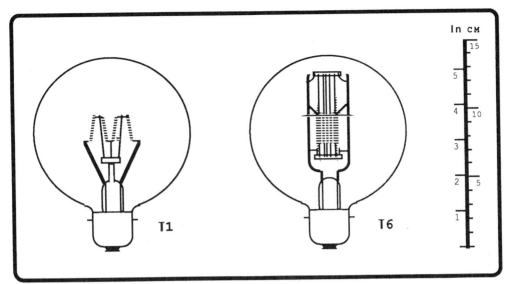

Figure A.17 Prefocus lamps

smaller, and constructed from stronger quartz glass. The high temperature filaments in these lamps also emit a greater amount of light. Thus tungsten halogen lamps, being smaller, brighter and longer lasting than their predecessors, have had a radical effect on lantern design, principally in regard to size.

Some quartz halogen lamps are shown in Figure A.18. Note the similarity of construction. All have a **ceramic bi-pin base**, a **flat filament**, and a small, flattened envelope. The **T25** on the left is a 500 watt lamp, used in small lanterns like the **minim fresnel** (Fig. 3.18d, p.23). In the centre is the **T19**, a 1000 watt version used in a number of different lanterns, including the **Harmony fresnel** (Fig. 3.18c, p.23) and the **LSC Starlette PC** (Fig. 3.21b, p.24). On the right is the **T29**, rated at 1200 Watts, which is used in the recently developed Strand 'Cantata' range (see Fig. 3.16, p.22 and Fig. 3.21a, p.24).

    Tungsten halogen lamps do, however, suffer some disadvantages. Firstly, they cannot be touched with bare hands, since even newly washed fingers will deposit a small amount of oil on the glass. When switched on, this will weaken the envelope, which will bubble at the point of contact and possibly explode. For this reason, they are supplied in plastic envelopes, which should only be removed after the lamp has been inserted. (Touched lamps can, and should, be cleaned before use with a tissue or cloth dampened with methylated spirit.) Secondly, because their filaments operate at higher temperatures than the older lamps, tungsten halogen lamps are extremely vulnerable when lit to sudden movement and, unless they have been warmed first, to rapid alterations in intensity. Ironically, therefore, although they should last a great deal longer than the older types, they often suffer an untimely accidental death at the hands of inexperienced or careless operators.

Since they are considerably more expensive than their predecessors, this irony is doubly painful to theatre managers.

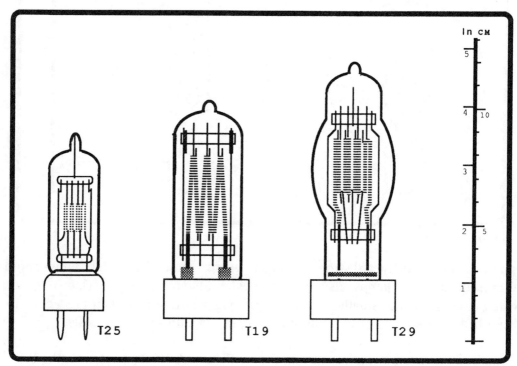

Figure A.18 Quartz halogen lamps

Reflectors

Three principal types of reflector are used in theatre lamps: **spherical**; **parabolic**; and **ellipsoidal.**

Spherical reflectors

The most useful characteristic of the **spherical** reflector is that, if a source of light is placed at its focal point (i.e. the centre of the sphere), all the light rays will be reflected back through the source. Thus the amount of light emitted from the source in one direction is vastly increased.

Fresnel and PC spotlights employ a spherical reflector, mounted in a unit which maintains a constant relationship between lamp and reflector.

Spherical reflectors are also used in some types of floodlight. With no lens system to focus the beam, the spherical reflector in this case produces a very wide distribution of light.

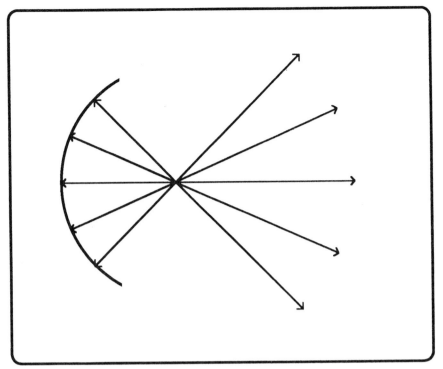

Figure A.19 Spherical reflector with light source at focal point

Parabolic reflectors

Light rays from a source placed at the focal point of a **parabolic** reflector will be directed away from the reflector in parallel lines. This results in a narrow, intense beam of light which will produce sharp-edged shadows. The most common use of this type of reflector is in car headlamps. In the theatre, of course, they are found in **beam lights** and **PAR cans.**

Some beam lights use a primary circular reflector to re-direct light from the lamp to the larger **parabolic** reflector. This also prevents unfocused light leaving the instrument directly from the lamp (see Fig. A.21).

Ellipsoidal reflectors

The **ellipsoidal** reflector is the most efficient of the three types. As Figure A.22 shows, the ellipsoidal reflector has two focal points. If a light source is placed at the **primary** focal point (within the reflector), then all the light reflected will pass through a second point, the **conjugate** focal point. This phenomenon is extremely useful, and is used to great advantage in the profile spotlight (also known, therefore, as the **ellipsoidal reflector spotlight**).

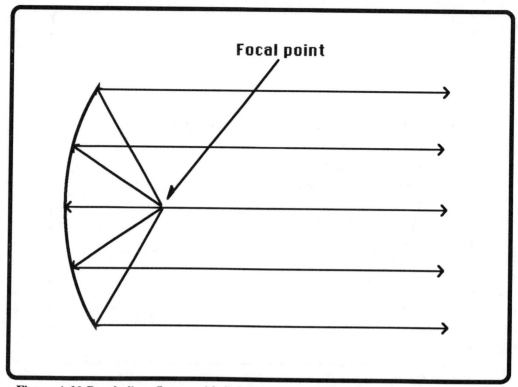

Figure A.20 Parabolic reflector with light source at focal point

By placing light-shaping devices such as shutters, irises, or gobos within the beam of light just before the conjugate focal point, and then focusing the beam with a lens or lenses, a sharply defined image of the required shape, many times magnified, can be projected on to a distant surface. Ellipsoidal reflectors are also found in some types of floodlight.

Lenses

Lenses used in theatre spotlights are almost invariably based on the plano-convex lens. As shown in Figure A.23, the light-bending characteristics of the plano-convex shape are most useful for concentrating a source of light into a beam of variable size.

The plano-convex lens is found in its simplest form in profile spotlights. Since diffusion of light is not a desirable characteristic, both surfaces of the lens are left clear. The disadvantages of the unmodified plano-convex lens are the unevenness of intensity within its field of light, and a tendency to produce rings of colour at the edge of the beam.

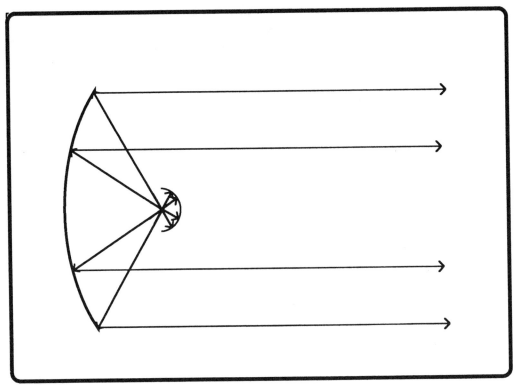

Figure A.21 Beam light with dual reflectors

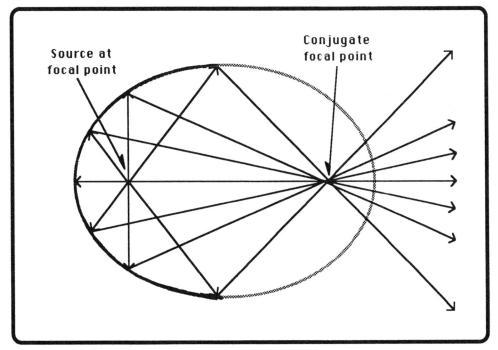

Figure A.22 Ellipsoidal reflector with light source at primary focal point

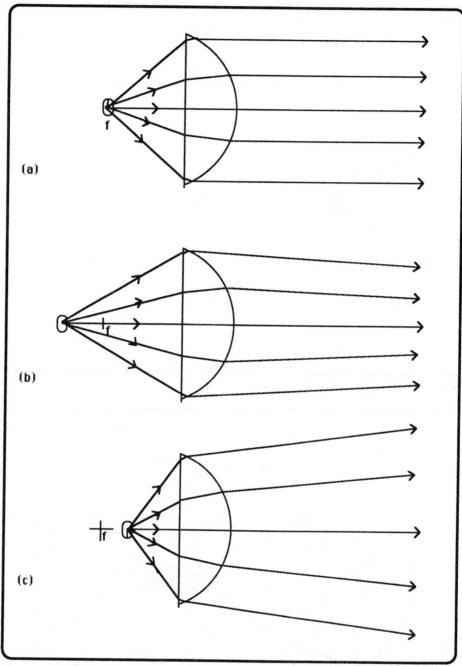

Figure A.23 Light bending properties of the plano-convex lens: (a) light source at the focal point; (b) source outside the focal point; (c) source inside the focal point

The pebble convex lens overcomes these problems. The plane surface of the lens is pebbled to diffuse the light, resulting in the more even intensity of light favoured for acting.

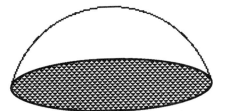

Figure A.24 Pebble convex lens

Though it may not be obvious at first inspection, the fresnel lens is also based on the plano-convex lens. Since it is the shape of the lens surfaces that determines how light will bend, rather than the thickness of the glass, the light-bending characteristics of the plano-convex lens have been preserved in a relatively flat lens by simply collapsing the convex surface down into a series of steps or rings. Frosting or pebbling of the plane surface adds diffusion, and masks the tendency of the fresnel to produce rings of varying intensity. The result is an extremely soft beam of even intensity, which is ideal for acting.

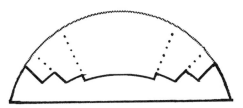

Figure A.25 Fresnel lens

Appendix III

Safety notes

Electrical safety

A note about electrocution, voltage and current

It is a popular misconception that fatal electrocution cannot occur with low voltages. Actually, fatalities have occurred at voltages as low as 24 volts ac. The danger in electrocution is not from the level of voltage, but from the amount of current that passes through the body, the path it takes, and its duration. Given a sufficient supply, and a good circuit through the body, a large current can flow from a low voltage.

Also, the level of current sufficient to cause injury or death is much lower than is generally appreciated. Currents as low as 1 – 6 milliamperes are sufficient to cause a painful shock, which may throw a person from a ladder and cause indirect injury. Above 6 milliamperes pain becomes intense and a person can become 'frozen' to a circuit. At 50 milliamperes, still a relatively small current, the heart can be thrown into fibrillation and death occur. Above 100 milliamperes, unless the duration of the shock is extremely short or its pathway through the body remote from major organs, death is a very real possibility. (As a guide to the current levels referred to, a small portable cassette radio draws 30 – 40 milliamperes, a household stereo system 120 – 160 milliamperes, an electric iron 5 amperes. 1 milliampere = 1/1000th of an ampere.)

Safety notes for lighting designers and operators

1. Unless you are qualified to do so, do not undertake any internal inspection, modification or repair of electrical or electronic equipment.

2. Never assume, because there are no visible signs of electrical activity (lamps burning, indicator lights on, etc.), that it is safe to proceed. The only safe assumption to make in regard to any electrical circuit is that it is **unsafe at any voltage. Always disconnect power before any inspection or work**.

3. Never undertake rigging and patching of lighting equipment unless the power supply to the dimmers is turned **off**.

4. When replacing burnt-out lamps in lighting equipment, first disconnect power supply to the lanterns. **N.B. Even with the relevant dimmer and master control set at zero on the lighting console, and the control key switched off, there may still be a dangerous voltage applied to the lantern. Before opening the lantern, disconnect it from the outlet.**

5. Always wear rubber-soled shoes when working on electrical equipment.

6. Do not use metal ladders for electrical work unless they are equipped with rubber feet.

7. When working on high ladders, always have another person present in case of accident.

8. When focusing lanterns, never use more than 70 per cent of dimmer power.

9. When hanging lanterns, always ensure that the power lead is clear of the body of the lantern, to avoid melting or burning of the lead and subsequent short-circuit.

10. When using portable power tools, use double-insulated equipment wherever possible.

11. If an overload occurs in any circuit, resulting in a fuse burning out or a circuit-breaker being tripped, always check equipment in the circuit before reconnecting. **If in any doubt, contact a qualified electrician before proceeding.**

Ladder safety

Ladders are extremely dangerous pieces of equipment unless they are handled correctly. Follow these instructions carefully, and you will live to climb again!

Step ladders

1. Always ensure that the chain, rope or metal arms limiting opening of the two sections are fully extended.

2. Point the ladder at the work, allowing sufficient space for you to work over the top.

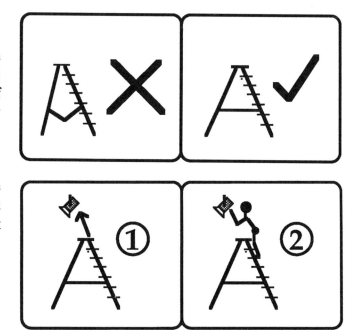

3. Do not stand on the top two rungs of the ladder. You cannot maintain balance effectively there.

4. Work over the top of the ladder, not out to the sides. The ladder is unstable otherwise.

5. Never leave tools or items of equipment on top of the ladder. You may forget they are there, with disastrous results.

6. When you move a ladder, look up to ensure you do not encounter overhead equipment.

Extension Ladders

When using an extension ladder, the distance between the wall and the foot of the ladder should be one third of the height of the ladder.

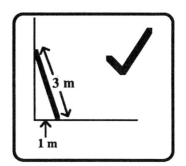

At lesser distances, the ladder may fall backwards.

At greater distances, it may slide down the wall.

Recommended books for further study

Bentham, Frederick, *The Art of Stage Lighting*, Theatre Art Books, New York, 1979.
A well-written book with a practical approach.

Carpenter, Mark, *Basic Stage Lighting*, NSW University Press, Sydney, 1982.

Palmer, Richard, *The Lighting Art*, Prentice-Hall, Englewood Cliffs, New Jersey, 1985.
An interesting book which deals mainly with the aesthetics of stage lighting.

Parker, Oran and R. Craig Wolf, *Scene Design and Stage Lighting*, Holt, Rinehard and Winston Inc, Fort Worth, 1990.
Up-to-date, comprehensive, and extremely well-written. Highly recommended, in spite of the expense.

Pilbrow, Richard, *Stage Lighting*, Nick Hern Books, London, 1992.
A popular book by one of Britain's best-known designers. Distributed in Australia by Currency Press, Sydney.

Index